Mines of *Churchill and Mineral Counties*

by
William O. Vanderburg

This publication is a reissue of U.S. Bureau of Mines Information Circulars 7093, originally issued in 1939, and 6941, originally issued in 1938.

Published by Stanley Paher

Nevada Publications
Box 15444
Las Vegas, Nevada 89114

Books in this series:

Mines of Churchill and Mineral Counties

Mines of Lander and Eureka Counties

Mines of Humboldt and Pershing Counties

Mines of Clark County

Mines of Goldfield, Bullfrog
 and other Sourthern Nevada Districts

Mines of the Silver Peak Range, Kawich Range,
 and other Southwestern Nevada Districts

Mines of Tuscarora, Cortez,
 and Other Northern Nevada Districts

Mines of Cherry Creek, Ely Range,
 and other Eastern Nevada Districts

Mines of Battle Mountain, Reese River, Aurora
 and other Western Nevada Districts.

Mining Districts and Mineral Resources of Nevada

Copyright © 1988
by Stanley W. Paher
ISBN 0-913814-73-3
PRINTED IN THE U.S.A.

FOREWORD

In Fall of 1936 and Fall of 1938 William O. Vanderburg issued information circulars on Churchill and Mineral counties in west central Nevada. Each was one of a series of circulars dealing with mining and milling operations in various mining districts in the Western States. All were published under the direction of the United States Bureau of Mines.

The author set forth data on operating costs, grades of ore treated, wage scales, haulage rates, and information on mining properties from existing literature, mine operators, and local sources during the course of field work. This reprint of these two information circulars gives us a picture of mining and milling operations in Churchill and Mineral counties at the time of original publication.

The information in this book is reliable as to conditions at the time author Vanderburg compiled the original reports. Though later developments have brought many changes, the publisher and other interested parties recognize the importance of these reports and decided to reissue them under the new title, *Mines of Churchill and Mineral Counties*. Three other books in this series give similar coverage of mines and mining in Humboldt, Pershing, Lander, Eureka and Clark counties (See back of book for details).

ACKNOWLEDGMENTS

The author wishes to thank the owner, lessees, and mine operators, too numerous to mention individually, who furnished aid and information during the course of the field work. Special thanks are due John T. Reid, Lovelock, Nev., who kindly furnished data of historical nature on the old mining districts in the Stillwater Range.

B.F. Couch, secretary of the Nevada State Bureau of Mines, Reno, Nev., made available published data pertaining to former mining activities within the county; credit for such data as were used is given in the text. Charles White Merrill, of the Mineral Production and Economics Division, Bureau of Mines, compiled the figures for the mineral-production tables for the Wonder and the Fairview districts, and for Churchill County.

TABLE OF CONTENTS

Map of Churchill County	8
Introduction to Churchill County	9
Churchill County General Information	
Political history, Topography, Water resources, Climate and vegetation, Power, Transportation, History of mining, Mineral production	11
Alpine District	20
Nevada gold group	20
Bernice District	21
Silver deposits	21
Antimony deposits	21
Chalk Mountain District	21
Chalk Mountain Silver Lead Mines Co.	22
Desert District	23
Manitou Gold Mining Co.	23
Eastgate District	23
Gold Ledge group	23
Buffalo Hump group	25
Building stone	26
Fairview District	26
Nevada Range Mines Co., Inc.	30
Gold Basin Mining Co.	31
Belle Mountain Mining Co.	31
Shamrock group	32
Westgate custom mill	32
Manganese deposit	32
Fireball District	32
Holy Cross District	33
Terrell group	33
Cinnabar Hill group	34
Bimetal group	34
Diatomaceous earth deposit	35
I.S.L. District	35
Black Prince group	35
Gold Bar group	36
Revenue group	37
Jessup District	37
Valley King group	37
Diatomaceous earth	38
Lake District	38
Limestone deposit	38
White Plains salt deposit	39
Sodium nitrate deposit	39
La Plata District	39
La Plata mine	40
Michigan Claim	40
Leete District	40
Sand Springs District	41
Dan Tucker mine	41
Salt depostis	42
Borax deposits	43
Shady Run District	43

Soda Lake District	44
Table Mountain District	43
Nickel-cobalt depostis	45
Dixie Comstock mine	47
Kaolin deposit	47
Dixie Marsh	48
Other Mines and prospects	48
Toy District	49
Toy mine	50
Hardscrabble claim	50
White Cloud District	51
Nevada United Mining Co.	51
Wonder District	52
Nevada Wonder mine	52
Churchill County footnotes	58
Mineral County General Information	
Topography, Water resources, Climate and vegetation,	
Power, Transportation, History of mining, Metal production	59
Map of Mineral County	63
Ashby District	64
Ashby Gold Mine, Inc.	64
Aurora District	64
Basalt District	71
Somerville group	71
Diatom Company	71
Bell District	71
Simon Silver-Lead Mines, Inc.	72
Omco mine	72
Golden Mile group	73
Clay Peters group	74
Harvey-Taylor group	75
Finger Rock Quicksilver Mining Co., Inc.	75
Diatomaceous earth	75
Broken Hills District	76
Broken Hills mine	76
Silver Trailer group	76
Baxter mine	77
Candelaria District	77
Argentum Mining Co.	78
Secretary Lode Mines Co.	78
Turquoise and variscite	80
Double Springs Marsh District	80
Eagleville District	81
Highland group	81
Other claims	82
Fitting District	82
Dover group	82
Other andalusite claims	83
Chiatovich group	83
Mica, Graphite, Iron	83
Fick Mining Co.	84
Hawaiian group	84
Garfield District	85

Hawthorne District	85
Lucky Boy Consolidated Mines Co.	85
La Panta mine	85
Pamlico mine	88
Placer gold, Barite	89
King District	89
Donnelly group	89
Marietta District	89
Moho mine	90
Endowment mine	91
Rutty group	92
Gold Gulch Mining and Milling Co.	92
Annett group	92
Mountain View District	92
Mountain Grant District	93
Grant Mountain Gold Mine	93
Big Indian mine, Cory mine	94
Talisman group, Return group, Molybdenite	94
Mount Montgomery District	95
Mount Montgomery Quicksilver Co.	95
Tip Top mine	95
Golden Gate Mining Co.	95
Mogoe Claims, Bentonite, Fluorspar	96
Oneota District	96
Pilot Mountains District	97
Mina Mercury Co.	98
Drew mine	98
Other cinnabar claims	99
Gunmetal group	99
Stormland group	100
Belleville mine	100
Sodaville tailings, Montezuma mine, Bentonite	101
Rand District	101
Randall property	101
Gold Pen mine	103
Lone Star group	103
Rawhide District	104
Leonard lease	104
Placers	108
Tungsten claims	109
Rhodes Marsh District	109
Santa Fe District	111
New Year group	111
Dolly group	114
American Copper Co.	115
Silver Star District	115
General Tungsten Corp, Silver Dyke mine	118
Tungsten Dike Group	121
Nevada Douglas Gold Mines, Inc.	121
High Ore group	121
Bentonite	122
Teel's Marsh District	122
Whiskey Flat District	123
Mineral County Footnotes	124

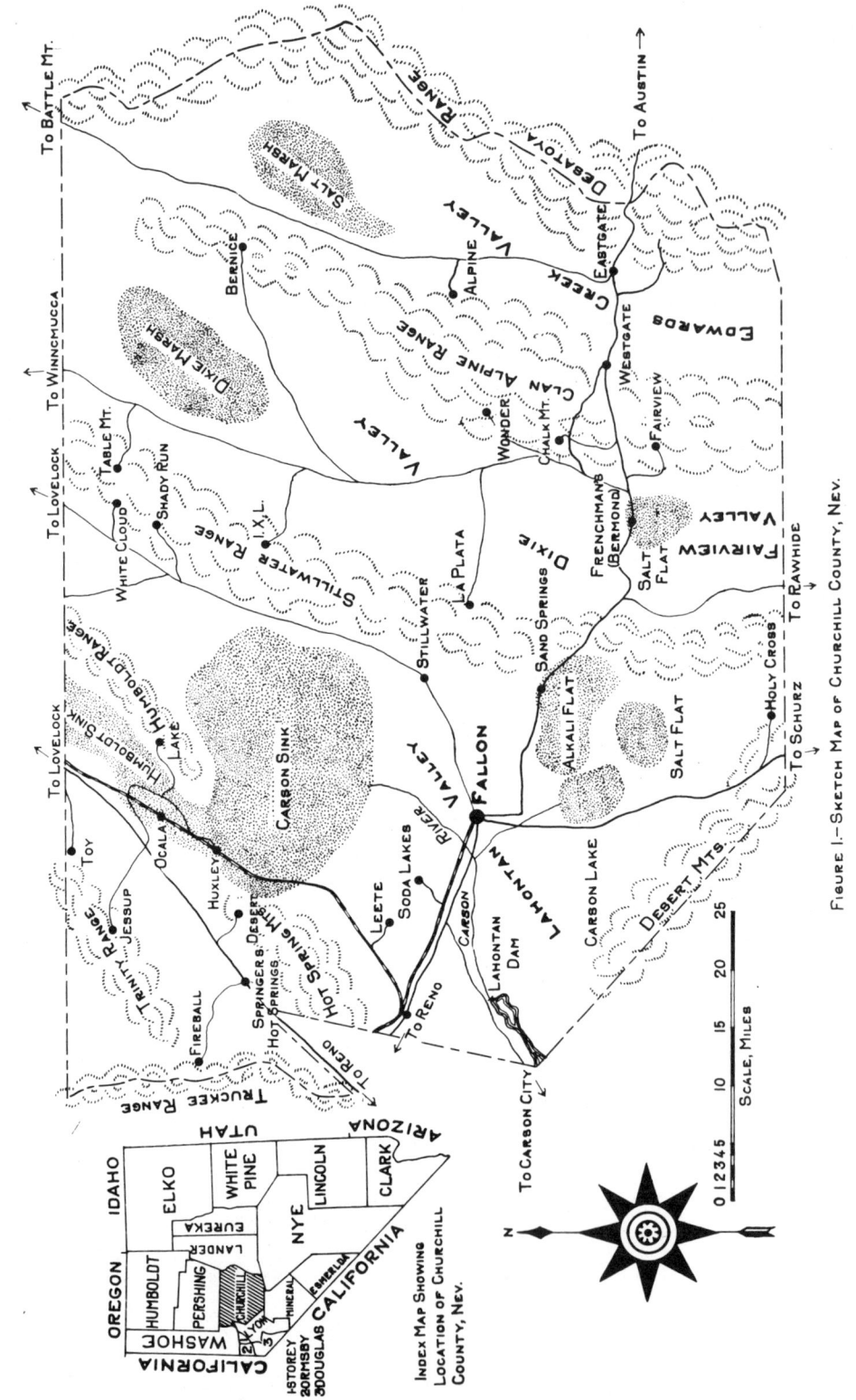

Figure 1.—Sketch Map of Churchill County, Nev.

INTRODUCTION TO CHURCHILL COUNTY

This report gives the results of a reconnaissance of the mining districts in Churchill County, Nev., made in the fall of 1938 and spring of 1939. All of the mining districts in the county were visited, but no attempt has been made to include all the mines and prospects within the area; inclusion or omission of a property in this paper has no bearing on its merits. The report gives the situation of the various districts, mineral production, and ownership of properties; describes the ore deposits, mines, and prospects; contains information on past and current activity; and includes other data likely to be useful to those interested in the development of our mineral resources.

The various districts were named when they were organized for purposes of record and regulation. When established, the districts embraced large areas without well-defined borders, so that the names have little geographical significance.

As a result of the mining activity in the Comstock and Reese River districts in Nevada, considerable prospecting in Churchill County took place in the 1860's and 1870's. A number of districts were organized[1], but no outstanding metal discoveries were made until 1905 and 1906, when rich silver-gold deposits at Fairview and Wonder were found. The Nevada Wonder mine in the Wonder district and the Nevada Hills mine in the Fairview district have been the principal producers in the county. With exhaustion of ores in these two properties about 1920, mining activity declined. Current mining activity consists largely of the operations of lessees; the ores produced are shipped either to smelters in Salt Lake Valley, Utah, and McGill, Nev., or to a local custom milling plant recently erected at Westgate in the southeastern part of the county.

In the industrial mineral group, salt for metallurgical use was an important commodity produced from the salines in Churchill County in the past. Sodium carbonate and borax also were produced in commercial quantities in former years.

Massive stone and rock ruins of the 10-stamp mill at Clan Alpine have withstood 120 years of weather and vandals. But the district's silver veins were slight, and this mill saw less than three years of use.

CHURCHILL COUNTY GENERAL INFORMATION

Political History

The region within the present boundaries of Nevada originally comprised part of the territory acquired by the United States from Mexico under the treaty of Guadalupe Hidalgo signed in 1848. After its acquisition by the United States, it constituted the western part of the Territory of Utah, from which it was separated by an Act of Congress approved March 2, 1861, to form the Territory of Nevada. The territorial status was maintained until October 31, 1864, when, by proclamation of President Lincoln, it was admitted into the Union as the thirty-sixth State.

Churchill County is one of the nine original counties created by Territorial Act approved November 25, 1861.[2] As created, the county was considerably larger; its size was reduced by cessions of land to Lander County on December 12, 1862; to Lyon County, February 20, 1864; and to Nye County, March 5, 1869. In February 1869 a small triangular tract was withdrawn from Humboldt County and ceded to Churchill County, making the boundaries as shown in figure 1.

When Churchill County was created, it was officially attached to Lyon County for judicial and revenue purposes, and a joint county seat was established at Buckland Station (later Fort Churchill). On February 20, 1864, Churchill County was separated from Lyon County, and in April of the same year La Plata, on the east side of the Stillwater Range, was designated the county seat. In 1868 the county seat was moved to Stillwater, on the west side of the range 18 miles west of La Plata, where it remained until 1908, when it was moved to Fallon.[3]

Churchill County derives its name from Fort Churchill, United States military post from 1860 to 1870, situated in Lyon County west of the present county line. It has an area of 5,091 square miles and a population of 5,075, according to the census of 1930. Most of the population resides in Fallon, the county seat and supply center for the ranches and mining districts within a radius of many miles.

The assessed value of real property in the county for the fiscal year ended June 30, 1938, was $8,108,312, and the county tax rate for the same period was $2.74 per $100, exclusive of special taxes, but including the State tax of $0.73 per $100 valuation.[4]

Topography

Nevada is in the Great Basin and lies between the Sierra Nevadas on the west and the Wasatch range on the east. The northern part of the Great Basin, in which Churchill County is situated, has a general altitude of about 4,000 feet and is traversed by long, narrow mountain ranges having a general north-south trend. The mountain ranges are separated by valleys ranging from 5 to 20 miles in width.

The Great Basin is an area of interior drainage; one of its peculiar topographical features is the number of playa lakes that occupy the lowest portions of the valleys. Churchill County contains a greater number of such lakes in proportion to its size than any other county in the State. They comprise the Humboldt Sink which receives the drainage of the Humboldt River from the east; the Carson Sink, which is the natural drainage basin for the Carson River flowing from the west; and the Dixie Sand Springs and Edwards Creek salt marshes, which drain valleys of the same names. During periods of precipitation they are covered with shallow sheets of water which may extend for many square miles, but during the summer months they evaporate to dryness, their beds becoming level mud plains so compact and hard that the imprints of passing automobiles are barely discernible. The beds consist of a mixture of alkali salts, clay, and silt. the salts generally present are sodium chloride, sodium carbonate, sodium bicarbonate, and sulfate of soda, with occasional small amounts of potassium chloride and borates. In the early days of silver milling by the Washoe and Reese River processes, the playa lakes were of considerable economic importance as a source of salt for chloridizing silver ores. The playas were also important as a source of borates in former years.

The principal mountain ranges in Churchill County, from east to west, are the Desatoya Range on the border between Churchill and Lander Counties, Clan Alpine Range, Stillwater Range, and Hot Springs Range. Other minor ranges are in the western part of the county. The higher ranges generally are rugged and form serrated crests having an altitude of 1,000 to 5,500 feet above the subjacent valley plains. The alternate arrangement of valleys and mountain ranges is the result of orographic displacement, the uplifted portions of the fault blocks appearing as long, narrow ranges.

The principal valleys in the county, from east to west, are Edwards Creek, Fairview, Dixie, and Lahontan. Dixie Valley, lying between the Stillwater and Clan Alpine Ranges, is unusual in that the floor of the valley is nearly 500 feet lower than that of the other valleys in the county.

Water Resources

Two of the largest rivers in Nevada normally discharge into sinks within the borders of Churchill County. The Humboldt River, with its tributaries, constitutes the largest stream system in the State; it rises in the northeastern corner of Nevada and, after flowing in a general west-southwest direction for at least 400 miles, finally empties into Humboldt Sink, part of which is in northern Churchill County. In 1935 the Federal Government constructed the Rye Patch Dam northwest of Lovelock to impound the waters of the Humboldt River; the entire flow is used for irrigation in the valley south of Lovelock, so that no waters reach Humboldt Sink.

The Carson River rises on the eastern slope of the Sierra Nevadas in California and flows northerly and easterly, entering Nevada as two separate streams known as East Fork and West Fork, which unite a few miles from the State line to form the main river. The Carson River normally empties into Carson Sink, but its waters are impounded in Lahontan Reservoir, which supplies the Newlands reclamation project, the first Federal irrigation project in the west. The water is used to generate hydroelectric power and irrigate farm lands in the vicinity of Fallon.

The Stillwater and Alpine Ranges are traversed by V-shaped canyons in which flow small mountain streams, several of which are perennial. Artesian water is present in Dixie and Edwards Creek Valleys. Hot springs occur in Dixie Valley on the west side of Dixie Salt Marsh and at Hot Springs station on the west side of Hot Springs Mountains.

The valleys between the mountain ranges are the natural underground drainage receptacles for the run-off, so that, if not otherwise available, water for mining and milling purposes can be developed by sinking wells, usually within convenient distance of most of the districts mentioned in this report.

Climate and Vegetation

The amount of precipitation in Churchill County varies progressively with the altitude, the mountains receiving the most and the valleys the least. Data compiled by the Nevada Agricultural Experiment Station indicate that approximately 46 percent of the area has an annual precipitation of 0 to 5 inches, 27 percent, 5 to 8 inches; 24 percent, 8 to 15 inches; and the remaining 3 percent, comprising the higher mountain ranges in the eastern part of the county, 15 to 20 inches. The precipitation is largely in the form of snow. With the exception of some of the more isolated areas within the county that may be snowed in for brief periods, all parts of the county are accessible the year around.

The climate is moderate and healthful and marked especially during the summer months by a large diurnal range in temperature, warm days being followed by cool nights. During the summer the days are warm, and there is an abundance of sunshine. The seasons are not distinct, and the transition from one to the other is almost imperceptible. In the spring and fall, winds rise with considerable regularity each day and form whirlwinds, which, on the dry lakes, whirl clouds of sand and alkali dust high into the air, often to a height of 1,000 feet.

Sagebrush, which grows throughout the county except on the higher mountain slopes above a general altitude of 6,500 feet is the dominant plant growth in the county. The saline marshes are sterile except for sparse growth of salt-loving plants, such as salt grass and tule, along their borders. The sagebrush lands are generally arable as they contain all the minerals necessary for raising crops adapted to the temperate zone. The larger portion of such land however, is unproductive, as rainfall is insufficient to support cultivated crops. Where water for irrigation is available, good crops are raised. The fertility of the soil is indicated by the size and thriftiness of the sage-brush, high growths indicating a rich soil containing more than the ordinary amount of moisture. The tenumbi bush (so-called by the Indians) is abundant in Dixie Valley and appears to be associated with shallow ground water, to which the roots penetrate to obtain moisture.

On the mountains above an altitude of about 7,000 feet the greater precipitation favors sparse growths, such as dwarfed juniper (*Junipero occidentalis*), mountain mahogany (*Cerocarpus laediforius*), and pinon pine (*Pinus monophylla*) being the most common. The forest growths are suitable only for firewood.

White sage, bunch grass, and other varieties of nutritious grasses furnish forage for cattle and sheep.

Power

The only public service power company operating in Churchill County is in the Truckee-Carson irrigation district, which serves Wadsworth and Nixon in Washoe County, Fernley and Fallon in Churchill County, and the surrounding rural populations from the Lahontan hydroelectric plant 17 miles west of Fallon. The plant has a capacity of 1,500 kilowatts supplied by three hydro-electric generators; the current is distributed through 283 miles of transmission and distribution lines at 33,000, 11,950, and 6,900 volts. None of the mining districts mentioned in this report is served by the district. Power for mining and milling is supplied by internal-combustion engines.

Transportation

The main line of the Southern Pacific R. R. traverses the northeast portion of Churchill County, and a branch line 16 miles in length connects Hazen on the main line with Fallon, the county seat.

The freight rates on ore and concentrates per ton from Fallon, the principal shipping point within the county, to Utah smelters are as follows:[5]

Value per ton	$15	$20	$30	$40	$50	$60	$70
40-ton minimum	3.30	3.30	4.00	4.70	5.40	6.10	6.80
20-ton minimum	5.20	6.00	6.50	7.00	7.70	8.20	8.70

Value per ton	$80	$90	$100	$150	$200	$250	$300
40-ton minimum	7.50	8.20	8.90				
20-ton minimum	9.20	9.70	10.20	11.50	12.20	14.30	14.30

Fallon is also on U. S. Route 50, (Lincoln Highway) which passes through the county in a northwest-southeast direction. U. S. Route 40 (Victor Highway) cuts across the northwestern part of the county. Both of these roads are hard-surfaced. Unimproved branch roads connect with the main highways at various points, so that virtually all of the districts mentioned in this report are accessible by automobile.

History of Mining

Churchill County was inhabited by Indians long before the appearance of the white men, but there is virtually no evidence to indicate that the aborigines were versed in the art of mining. Vestiges of prehistoric Indian occupation are found in the form of petroglyphs chiseled in caves and on ledges in several localities and in articles of Indian workmanship found in several caves in the northern part of the county. The mysterious petroglyphs were chiseled by people who have loft no decipherable records of their origin or state of civilization. Some of these rock inscriptions are found west of the Fallon Schurz highway, 12 miles southwest of Fallon, along terraces left by the receding water of ancient Lake Lahontan, which at one time covered an area of about 8,000 square miles.

In the northern part of Churchill County are several limestone caves that were used as dwellings by Indians. The largest of these caves, situated 22 miles southwest of Lovelock, was located as a mining claim in 1911 by James H. Hart and David Pugh, of Lovelock, Nev., for the bat guano found therein. During the removal of about 250 tons of guano for fertilizer, numerous Indian artifacts were found, and subsequently the cave was explored by archaeologists. Harrington[6] states it is probable that articles found in the debris range in age from 1,000 B. C. to recent time. It is interesting to note that of about 10,000 specimens of ancient Indian civilization unearthed in the cave, no metals were found.

The mining history of Nevada can be divided into two periods of extraordinary activity. The first began in the sixties with the discovery of the Comstock lode and continued until about 1885: the second period began with the discovery of Tonopah at the beginning of this century and continued for about 3 decades.

Although Churchill County was traversed by the two main pioneer overland travel routes across the State taken by emigrants to California in the gold rush of '49 and the years following, no metal was discovered until the early sixties. Humboldt and Carson Sinks, in the western part of the county, known to the pioneers as the Forty-Mile Desert, constituted one of the most formidable barriers to early-day travel, and the country appeared so forbidding that there was little to induce them to tarry or to invite their return.

The first recorded mineral discovery in the county was made by Asa L. Kenyon in 1855, when soda was found in the Soda Lakes near Ragtown. Kenyon, the first settler in the county, established a trading post at Ragtown in 1854. Ragtown (afterward called Leeteville) is one of the historical sites in Nevada, since it was here that the emigrants recruited their stocks and rested after crossing Humboldt and Carson Sinks. It was a station on both the Humboldt River route and the Simpson route; the latter route, laid out by Capt. (later Col.) J. H. Simpson for the United States Government in 1859, was used by the pony express and later by the Overland Mail & Stage Co., and was the principal line of travel across the State until the completion of the Central Pacific R. R. (now Southern Pacific R. R.) in 1869. This route went by way of La Plata, New Pass, Jacobsville, and a point near Austin, coming within a few miles of the two most important mineral discoveries in Churchill County, Fairview and Wonder; but, notwithstanding this fact, the mining camps at these places were not established until nearly half a century later. Although Churchill County was overrun by prospectors in the sixties, no important mineral discoveries were made until many years later.

The status of the metal-mining industry in Churchill County in the late sixties is given in the following extract from a Government report by Browne and Taylor.[7]

In Churchill County there are three districts that have attracted some notice, because of the supposed valuable ledges they contained. These are severally named the Silver Hill, the Mountain Well (La Plata), and the Clan Alpine, and to them most of the work performed in the county has been confined. There are in this county four quartz mills carrying 55 stamps, and having driving power equal to that of 165 horses. The total cost of these mills was $395,000. Three of them are at Mountain Well, and one not quite finished in the Clan Alpine district. They have produced but a few thousand dollars: worth of bullion all told, none of them having been able to run for more than a few days at a time, from an insufficient supply of pay ore, but few of the ledges here having been opened up to even the superficial depths common to most other districts. In the higher strata of some of them small aggregations of very rich ores have been found, and the chances favor the supposition that when properly developed they will afford enough ore to keep. the present and perhaps additional mills running. Very few additional mills, however, can be operated in the western half of the county, owing to the limited supply of wood and water.

The production of salt from the dry lakes in Churchill County attained considerable importance in the sixties and seventies. The salt was largely used in reducing silver ores on the Comstock and other silver camps in the northern part of the State.

With the discovery of Tonopah in Nye and Esmeralda Counties in 1901 a revival of mining took place. All parts of the State were overrun by prospectors; and, as a result, a number of gold and silver deposits were found and new camps established. Churchill County was not overlooked, and during this period the deposits at Fairview (1905), Wonder (1906), and several others of less importance were found and worked. During this period speculation in mining was prevalent, and an abundance of money was available for mining ventures; miners obtained valuable leases and money for equipment and development with little difficulty; claim owners were able :o sell their holdings for cash with few or no showings. One of the striking developments in mining activity during the period was the widespread application of the leasing system whereby a claim or portion of a claim is worked upon payment of a royalty or certain percentage of the value of the ore produced. To the leasing system may be attributed the discovery of many camps in the State, including Fairview and Wonder.

After 1920, with exhaustion of the ore reserves at Wonder and Fairview, mining declined, so that for a number of years the only activity has been a small amount of desultory leasing operations or small company operations, the latter generally short-lived.

In the first part of 1939 the writer estimated that 50 men, mostly lessees, were employed directly in the mining industry in Churchill County.

Mineral Production

Statistics on mineral production of Churchill County before 1903 are incomplete. The net annual production of gold and silver from 1885 to 1903, as compiled from the reports of Churchill County assessors, is shown in table 1.[8] The figures in the table are based on the value of the ore after mining and milling expenses were deducted. The gross production for the period stated is estimated to have been about $250,000. From fragmentary data, the production of metals before 1885 is estimated to have been an additional $50,000, making a total of $300,000 up to 1903. This production was derived from a number of small properties in widely scattered districts.

With the discovery of the camps of Fairview and Wonder in 1905 and 1906, respectively, metal production increased considerably. From 1904 to 1937 the metal mined, chiefly silver and gold, had a total value of $10,330,698 as shown in table 2.

Table 1. Net annual production of gold and silver form Churchill County, 1185-1903
(compiled from quarterly assessment rolls of the county assessors)

Year		Gold and Silver	
		Tons	Values
1885		1,521	$57,380
1886			
1887		243	14,096
1888			
1889		200	12,000

Year	Value of gold	Value of silver	Total value
1890	$4,000	$6,000	10,000
1891	3,000	5,000	8,000
1892	2,000	2,000	4,000
1893	2,000	2,000	4,000
1894	[1]		
1895	2,000	5,000	7,000
1896	2,500	4,500	7,000
1897	1,500	4,000	5,500
1898	1,500	4,000	5,500
1899	1,800	4,500	6,300
1900	[1]		
1901	2,000	3,000	5,000
1902	5,000	8,000	13,000
1903	5,000	8,145	13,145
Total	32,300	56,145	171,921

[1] No figures given

Small amounts of nickel, antimony, and tungsten ores also have been produced. In the industrial-mineral group considerable quantities of salt, probably about 500,000 tons, have been mined, chiefly for metallurgical use. Between 1870 and 1873 borax was produced from the salines in Sand Springs Marsh, Dixie Marsh, and Soda Lakes. Although no statistics are available as to the amount produced, it has probably been less than 1,000 tons. The production of sodium carbonate from Soda Lakes between 1868 and 1893 averaged about 400 tons annually, making a production of about 10,000 tons for the period. In addition, small amounts of diatomaceous earth and shell limestone have been mined in the county in former years.

Current mining activity is confined mainly to leasing, and the production of shipping ore for the past 2 years has averaged about 300 tons per month.

I. C. 7093

TABLE 2.— Gold, silver, copper, and lead production in Churchill County, Nev., 1904-37, in terms of recovered metal
(Compiled by Charles White Merrill, Mineral Production and Economics Division, Bureau of Mines)

Year	No. of mines	Placer				Total Value
		Gold		Silver		
		Fine ounces	Value	Fine ounces	Value	
1904	—	—	—	—	—	—
1905-06	—	—	—	—	—	—
1907	1	—	—	—	—	—
1908	—	—	—	—	—	—
1909	—	—	—	—	—	—
1910	—	—	—	—	—	—
1911	—	—	—	—	—	—
1912	—	—	—	—	—	—
1913	—	—	—	—	—	—
1914	—	—	—	—	—	—
1915	—	—	—	—	—	—
1916	—	—	—	—	—	—
1917	—	—	—	—	—	—
1918	—	—	—	—	—	—
1919	—	—	—	—	—	—
1920	—	—	—	—	—	—
1921	—	—	—	—	—	—
1922	—	—	—	—	—	—
1923	—	12.24	$253	3	$2	255
1924	—	—	—	—	—	—
1925	—	—	—	—	—	—
1926	—	—	—	—	—	—
1927	—	—	—	—	—	—
1928	—	—	—	—	—	—
1929	—	—	—	—	—	—
1930	—	—	—	—	—	—
1931	—	—	—	—	—	—
1932	—	—	—	—	—	—
1933	—	—	—	—	—	—
1934	—	—	—	—	—	—
1935	—	—	—	—	—	—
1936	—	—	—	—	—	—
1937	—	—	—	—	—	—
Total	1	12.24	$253	3	$2	255

TABLE 2.— Gold, silver, copper, and lead production in Church-
hill County, Nev., 1904-37, in terms of recovered metal (cont'd.)
(Compiled by Charles White Merrill, Mineral Production and Economics
Division, Bureau of Mines)

Year	No. of mines	Lode						
		Ore	Gold		Silver		Copper	
		Short tons	Fine ounces	Value	Fine ounces	Value	Pounds	Value
1904....	3	460	306.72	$6,340	—	—	—	—
1905-06.	—	—	—	—	—	—	—	—
1907....	6	7,316	7,371.38	152,380	655,480	$432,617	—	—
1908....	16	2,090	2,656.76	54,920	265,425	140,675	5,212	$688
1909....	8	1,211	1,723.74	35,633	140,204	72,906	—	—
1910....	15	1,164	2,083.13	43,062	162,541	87,772	2,417	307
1911....	14	19,463	5,998.02	123,990	456,066	241,715	—	—
1912....	10	58,434	17,325.70	358,154	1,434,068	881,952	6,196	1,022
1913....	12	83,901	15,630.15	323,104	1,383,257	835,487	4,292	665
1914....	19	114,771	15,435.16	319,073	1,545,926	854,897	3,892	518
1915....	13	123,991	13,866.40	286,644	1,620,573	821,630	19,167	3,354
1916....	20	110,943	13,734.15	283,910	1,386,524	912,333	9,609	2,364
1917....	13	60,587	8,408.35	173,816	887,765	731,518	4,030	1,100
1918....	13	49,950	4,970.09	102,741	625,152	625,152	1,435	354
1919....	12	40,892	5,816.13	120,230	488,053	546,619	—	—
1920....	9	1,292	534.54	11,050	26,686	29,088	—	—
1921....	9	70	32.60	674	3,837	3,837	194	25
1922....	8	2,269	106.38	2,199	40,971	40,971	13	2
1923....	11	1,104	114.74	2,372	30,601	25,093	174	26
1924....	10	101	13.55	280	5,157	3,455	129	17
1925....	9	1,445	123.45	2,552	67,204	46,640	1,914	272
1926....	8	896	210.43	4,350	27,490	17,154	1,453	203
1927....	7	368	60.37	1,248	7,078	4,013	443	58
1928....	4	625	138.35	2,860	4,687	2,742	1,112	160
1929....	3	976	13.96	289	3,147	1,677	498	88
1930....	3	1,047	39.09	808	3,327	1,281	55	7
1931....	4	417	245.45	5,074	13,396	3,885	—	—
1932....	5	225	43.85	906	748	211	—	—
1933....	6	78	47.09	973	1,677	587	—	—
1934....	17	2,545	1,728.16	60,399	12,124	7,838	1,596	128
1935....	19	5,378	1,926.25	67,419	33,348	23,969	372	31
1936....	26	4,173	1,937.00	67,795	60,792	47,083	—	—
1937....	15	2,738	896.00	31,360	61,419	47,508	—	—
Total.	—	700,920	123,537.14	2,646,605	11,454,723	7,492,305	64,203	11,389

I. C. 7093

TABLE 2.- Gold, silver, copper, and lead production in Churchill County, Nov., 1904-37, in terms of recovered metal (cont'd.)
(Compiled by Charles White Merrill, Mineral Production and Economics Division, Bureau of Mines)

Year	Lode				Total Value (lode and placer)
	Lead		Total Value	Average recoverable value of ore 1/	
	Pounds	Value			
1904....	—	—	$6,340	$13.78	$6,340
1905-06.	—	—	—	—	—
1907....	—	—	584,997	79.96	584,997
1908....	1,690	$71	196,354	93.95	196,354
1909....	26,602	1,144	109,683	90.57	109,683
1910....	31,602	1,390	132,531	113.86	132,531
1911....	117	5	365,710	18.79	365,710
1912....	6,538	294	1,241,422	21.24	1,241,422
1913....	162	7	1,159,263	13.82	1,159,263
1914....	10,518	410	1,174,898	10.24	1,174,898
1915....	24,574	1,155	1,112,783	8.97	1,112,783
1916....	8,902	614	1,199,221	10.81	1,199,221
1917....	33,951	2,920	909,354	15.01	909,354
1918....	39,090	2,775	731,022	14.64	731,022
1919....	22,932	1,215	668,064	16.34	668,064
1920....	12,659	1,013	41,151	31.85	41,151
1921....	38,393	1,728	6,264	89.49	6,264
1922....	545	30	43,202	19.04	43,202
1923....	82,802	5,796	33,287	30.15	33,542
1924....	37,803	3,024	6,776	67.09	6,776
1925....	919,626	80,007	129,471	89.60	129,471
1926....	447,788	35,823	57,530	64.21	57,530
1927....	142,859	9,000	14,319	38.91	14,319
1928....	219,156	12,711	18,473	29.56	18,473
1929....	140,315	8,840	10,894	11.16	10,894
1930....	30,348	1,517	3,613	3.45	3,613
1931....	200	7	8,966	21.50	8,966
1932....	—	—	1,117	4.96	1,117
1933....	8,820	326	1,886	24.18	1,886
1934....	152,232	5,633	73,998	29.08	73,998
1935....	7,233	289	91,708	17.05	91,708
1936....	6,000	276	115,154	27.60	115,154
1937....	36,000	2,124	80,992	29.58	80,992
Total	2,489,457	180,144	10,330,443	14.74	10,330,698

1/ Not to be confused with average assay value of ore.

ALPINE DISTRICT

The Alpine District is in eastern Churchill County on the east side of the Alpine Range, about 13 miles northwest of Eastgate. It can be reached by automobile from Eastgate over a fair desert road through Edwards Creek Valley. The district was originally organized as the Clan Alpine district in January 1864 but is now generally known as Alpine. In 1864 a number of claims were located for silver in the vicinity of Florence Canyon, and considerable work was done on them. In 1866 the Silver Lode Mining Co. erected a 10-stamp mill at the mouth of Cherry Creek, where the camp of Alpine was laid out. A number of stone buildings, now in ruins, attest this early activity. The veins proved to be narrow and low-grade, so that the district was abandoned after several years. Except for sporadic prospecting, there has been little mining since the early days. In all probability not more than a few thousand dollars has been produced. In April 1939 the only mining in this area was on the Nevada Gold group of claims.

Nevada Gold Group

The Nevada Gold group, comprising three unpatented claims owned by Sam Spring of Fallon, is at the head of Stone Canyon, 33 miles by road north of Eastgate. This property was prospected in the early days, but there is no record of any production. From the size of several small stopes, however, production probably has been several hundred tons of shipping ore. Early in 1939, Spring, working single-handed, mined about 50 tons of shipping ore.

Development consists of a crosscut adit 500 feet long that is filled with detrital material washed in by cloudbursts. A second adit has been driven on the vein for about 100 feet. These workings, with several scattered open-cuts and a shaft 130 feet deep, total approximately 800 feet. The only equipment on the property consists of tools for hand-mining and several camp buildings. There is a small perennial stream in Stone Canyon that probably could furnish enough water for a small mill.

The formation is slate intruded by monzonite. The ore occurs in two nearly parallel fissure veins traversing both the slate and monzonite. the vein on which most of the work has been done strikes N. 10° E. and dips 50° westerly. The economic minerals are silver, gold tetrahedrite, galena, and sphalerite and their oxidation products in a gangue composed chiefly of quartz with a little pyrite. Width of veins ranges from a few inches to 5 feet, and where the ore has been stoped the veins average about 2 feet in width.

BERNICE DISTRICT

The Bernice district is on the west slope of Clan Alpine Range in east Churchill County, 90 miles by road, a little south of west from Fallon and 70 miles southeast of Lovelock, both towns being on the Southern Pacific R. R. From Fallon the district can bs reached by automobile over the Lincoln Highway to Dixie Valley turn off, a distance of about 40 miles, and thence up Dixie Valley to the mines situated in Bernice and Hoyt Canyons. The last 25 miles of road is in poor condition, but with careful driving is passable.

Silver and antimony ores were discovered in the district in the late eighties. The principal ore mined has been silver ore from the Williams mine; probably about 500 tons of antimony ore has been shipped. In 1939 the only activity in this area was by lessees on the Bluebird group of claims.

Silver Deposits

The Williams mine at the head of Bernice Canyon comprises two patented and three unpatented claims owned by the Warren W. Williams estate. This property was discovered by James Wardell in the late seventies and was sold to Williams, who operated it from 1880 to 1890. A 10-stamp mill was erected below the mine in Bernice Canyon; silver produced is said to have amounted to $300,000. The ores contained considerable arsenic in the form of arsenopyrite, and the mill was equipped with two White-Howell roasters. The ore is in a series of narrow veins in shale. The outcrops carried considerable antimony, which decreased at depth. The mine has been inactive for many years.

The tailings pile below the old mill site, containing about 6,000 tons, is owned by C. M. Beeghly and L. T. Ellis, according to a location notice dated March 18, 1939.

The Bluebird group, comprising six unpatented claims owned by Albert Lofthouse, of Fallon, is in the northern part of the district. In 1939 the property was under lease to D. G. Brunner, A. J. Cooley, and associates, who, up to April 1939, had made one small shipment of ore to the McGill smelter. Here the ore occurs in a narrow vein in limestone.

Antimony Deposits

Antimony was first discovered in the district in the eighties by H. Hoyt. W. W. van Reed shipped several cars of hand-sorted ore containing 62 percent antimony to the Star and Matthews smelter in San Francisco. A small quantity of antimony ore was produced in 1893 by Sanders and Young, and the last production of record was made in 1906, when 200 tons were shipped. C. Solomon, Jr., of San Francisco, identified with the Chapman Smelting Co., was active in the district during the World War. Although considerable development work was done at that time there is no record of any production.

The Antimony King and Lofthouse mines are the principal antimony properties. The former is at Bernice and the latter about 5 miles southwest. The deposits have been prospected by numerous superficial markings. According to Mallory,[9] the prevailing geological formations are a series of sedimentary rocks, chiefly slate, shale, and limestone intruded by granite. The commercial antimony ore occurs in quartz veins in the lower slate stratum of the sedimentary series closely associated with limestone strata. At the Antimony King mine are two siliceous limestone beds averaging about 2 feet in thickness and being 300 feet apart. The most persistent bed is traceable on the surface for 3 miles. The croppings of limestone are stained yellow from antimony oxide along the cleavage planes, but it is not present in commercial quantities.

The Antimony King vein is traceable on the surface for 900 feet and strikes N. 10° E. with a dip of 50° W. It cuts the slate and limestone strata at nearly right angles. The width of the vein ranges from 4 inches to 4 feet, and both walls have a parting of black slaty gouge. The vein filling is predominantly quartz containing shoots of nearly solid antimony in places. The weathered zone contains cervantite, the yellow oxide of antimony, which in places forms coatings on stibnite. Sphalerite occurs sparingly. the best ore is near the intersection of the vein with the southern limestone stratum.

The Lofthouse vein parallels the Antimony King vein; the best ore likewise occurs associated with a limestone stratum in the enclosing shale. In the Lofthouse vein the antimony is less massive, occurring in fibrous or needle-like forms.

CHALK MOUNTAIN DISTRICT

The Chalk Mountain district is in the southeastern part of Churchill County in the vicinity of

Chalk Mountain in the southern end of Dixie Valley. This mountain, isolated from the main Alpine Range, is 3 miles long and 2 miles wide and rises to a height of about 1,000 feet above the surrounding terrain. It is composed of whitish dolomitic limestone, which forms a conspicuous land-mark in the region. The nearest railroad connection is at Fallon, 44 miles west-northwest of the principal mine. The district is midway between Wonder and Fairview, sometimes being considered part of the latter, which lies about 8 miles south. The mineral production is included under the Fairview district.

Chalk Mountain was prospected in the early years of mining activity in Churchill County, but received little attention until 1923, when E. M. Dawes interested a Minneapolis group and acquired control of the Chalk Mountain Silver Lead Mines Co. Dawes and associates started a development campaign and several deposits of ore were found. This development and increased production stimulated activity, and about eight small companies were organized, which were active during 1925 and 1926. These small companies produced little, and all became inactive after short periods of prospecting. In the early part of 1939 the only mining activity was by lessees on the property of the Chalk Mountain Silver Lead Mines Company.

Production of the district has totaled several hundred thousand dollars in shipping ore, virtually all of which was produced by the Chalk Mountain Silver Lead Mines Co.

Chalk Mountain Silver Lead Mines Co.

The Chalk Mountain Silver Lead Mines Co., E. M. Dawes, Lovelock, Nev., president, owns 12 unpatented claims on the east slope of Chalk Mountain. The company was incorporated in 1920 under the laws of Nevada, with a capitalization of 2,000,000 shares at a par value of 50 cents, which was later increased to $1 per share. The principal production was made between 1923 and 1929. In 1925, 1,505 tons of ore sold to a smelter had a value of $86,923 and contained 99 ounces of gold, 59,651 ounces of silver, and 861,355 pounds of lead. UP to June 1, 1927, the total production of the company amounted to 2,300 tons valued at $135,000, or an average net value of $58.70 per ton. Some ore was also shipped in 1928 and 1929. In 1929 a 50-ton mill employing table concentration and having a Diesel power plant was erected at the mine to treat the lower- grade ore, but it was unsuccessful metallurgically. In 1930 the company sold the mill equipment and suspended operations, since which time the mine has been operated sporadically by lessees with occasional small shipments. In 1939 the mine was under lease to Harry Howard and associates, who were shipping ore intermittently to the smelter. The royalty payments made by lessees on net smelter returns are 10 percent of the gross smelter value of ore assaying $35 or less, 20 percent on ore ranging between $35 and $75 per ton, and 25 percent on ore having a gross smelter value of more than $75 per ton.

The mine is developed by a 40-foot shaft, two 110-foot shafts, and one double-compartment vertical shaft 517 feet deep, with lateral work on 6 levels. Total workings comprise about 5,000 feet.

Equipment includes a 40-horsepower Fairbanks-Morse gasoline hoist, blacksmith shop, tools for hand mining, and camp accommodations for a crew of about 20 men.

Formation is dolomitic limestone intruded by granodiorite. The limestone is locally folded and traversed by a number of faults, with which the ore bodies are associated. Ore occurs as irregular replacement deposits ranging from 1 to 12 feet in width along fissures and bedding planes of the limestone. The ore minerals are cerussite, anglesite, cerargyrite, wulfenite, vanadinite, and argentiferous galena in a porous gangue consisting of quartz, calcite, altered limestone, and iron oxides. Ore is largely oxidized to the lowest level of the mine, and the high iron content makes it desirable for fluxing.

The mine dump, consisting of about 3,000 tons, is reported to be ore of milling grade.

DESERT DISTRICT

The Desert district is in the northeast portion of the Hot Springs Mountains in northwestern Churchill County on the northeast flank of Desert Peak, which has an altitude of 5,401 feet above sea level. Parran siding, on the Southern Pacific R. R., is 7 miles east. The district is accessible by automobile over a desert road 5 miles in length, which leaves the Victory Highway at a point several miles northeast of Springer's Hot Springs.

The principal mine, the Desert Queen, is said to have been discovered by emigrants in 1849. If this is true, it has the distinction of being the first lode mine worked in the northern part of Nevada. In the early sixties the Desert Queen mine was worked by a company from Schenectady, N. Y., with which Horace Greeley is reported to have been connected. In 1863, a mill consisting of two two-stamp batteries was built on the flat 2 miles east of the mine, but it proved unsuccessful and a second mill was built at the edge of Humboldt Sink 14 miles northeast. Eventually the property reverted to the Public Domain, and in 1931 a group from Lovelock, Nev., relocated the principal claims, which in the following year were taken over under bond and lease by Wilford Dennis and associates, who organized the Manitou Gold Mining Co. A 25-ton-capacity amalgamation-concentration mill was erected by the company in 1937. Early in 1939 a crew of three men was employed on development.

Manitou Gold Mining Co.

The Manitou Gold Mining Co. controls, under bond and lease, the Chrysler-Bonanza group of 10 unpatented claims owned by Herman N. Marker and K. 0. Olfers, of Lovelock, Nev.

Development consists of three adits, the longest 750 feet, and several shallow shafts and old workings, totaling approximately 4,000 feet. Mining equipment includes an Ingersoll Rand, two-stage compressor (model 25) with a Waukesha gasoline engine, a blacksmith shop, rock drills, and other mining tools. Mill equipment consists of a Blake-type crusher size 7 by 9 inches, a Straub ball mill, amalgamation plate, and an Overstrom concentrating table. Power is supplied by gasoline engines. Camp buildings can accommodate a crew of 10 men. Water for milling is pumped from a shaft 265 feet deep near the mill site.

Ore occurs in a series of fissure veins with a maximum width of 5 feet in a diorite formation. The principal vein, the Desert Queen, has a strike of N. 60° W. with a dip of 40° NE. The economic minerals, gold and silver, have a quartz gangue impregnated with iron oxides.

EASTGATE DISTRICT

The Eastgate district is on the west side of the Desatoya Range in southeastern Churchill County about 50 miles east-southeast of Fallon, the nearest railroad point. No information is available concerning the first mining activity in this area, although it is known that a number of gold and silver properties have been prospected in a desultory manner for at least 40 years. The production of the district is estimated at about $25,000 in shipping ore, most of which was derived from the Gold Ledge group of claims. In the first part of 1939 the only activity was at the Gold Ledge property.

Gold Ledge Group

The Gold Ledge group of unpatented claims, owned by W. H. Schweis, of Reno, Nev., is situated on a mountain spur off the west side of the Desatoya Range about 56 miles east southeast of Fallon, the nearest railroad point. It can be reached by automobile over an unimproved desert road 5 miles in length, which leaves the Lincoln Highway about 1 mile west of Eastgate. The last 2 miles of the road is in a narrow, steep canyon with a number of sharp turns. First locations in this area were made by E. W. Baker in 1906. The Gold Ledge property has been worked intermittently by individuals and small companies. In 1934 the property was operated for a short time by the Monarch Gold Ledge Mining Co., which erected a 50-ton amalgamation-concentration mill several miles northwest of the mine, but it was unsuccessful; the mill equipment was sold in 1939. About $20,000 in shipping ore has been produced.

Development consists of an adit 350 feet in length along the ledge, an inclined shaft 200 feet in depth with levels at 150 and 200 feet below the surface, a vertical shaft 110 feet deep, and other workings, totaling about 1,500 feet. Mining equipment includes a 7 by 6-inch Sullivan air compressor, belt-driven by an automobile engine; a West Coast, 5-horsepower, gasoline geared hoist; rock drills and other mining tools; a blacksmith shop; and camp accommodations for a crew of four men. Although there is no water at the mine, it can be obtained in Eastgate Wash, several miles northwest of the mine. In 1939 the property was being worked by Schweis with a crew of two men. In 1938 he shipped about 100 tons of ore, averaging approximately $30 per ton, to the Dayton custom mill at Silver City, Nev. The truck haul to the Dayton mill, a distance of about 160 miles over hard-surfaced roads, cost $5 per ton and milling was $4 per ton. The shipper was paid for 90 percent of the assay value of the ore.

Ore containing gold and silver occurs in a fault fissure vein striking N. 30° E. and dipping about 65° easterly. The formation is altered rhyolite. The vein ranges in width from 1 to 6 feet, with no well-defined walls. The gangue is composed largely of crushed and iron-stained rhyolite showing little silicification. At the south end of the workings the vein has been intercepted by a number of faults that cut it at right angles; to the north, a fault striking N. and S. and dipping 70° E. apparently has displaced the vein several hundred feet north, as indicated by workings on a vein east of the fault.

The smelter returns on a shipment of ore made by W. H. Schweis to International Smelting Co. at Tooele, Utah, on March 20, 1936, furnished the following data:

Metal quotations:	Gold	$35.00 per ounce.	
	Silver	.77 per ounce	
	Lead	4.6 cts. per lb.	
		Ounces	
Settlement assay:	Gold	.91	
	Silver	13.44	
		Percent	
	Copper	.07	
	Lead	2.0	
	Zinc	Nil	
	Sulfur	.5	
	Iron	1.8	
	Insoluble	90.2	
	Lime	.7	
		Pounds	
Net weight		107,500	
Moisture 2.64 per cent		2,838	
Dry weight		104,662	52.331 tons
Metal payment:	Lead, 50 percent at		
	$0.01075 per pound	$.215	
	Silver, 95 percent at 0.77 per ounce	9.831	
	Gold, 91 percent at $35 per ounce	28.984	
	Gross value per ton	39.030	

```
         Base treatment charge, per ton            $4.00
                 Charge, 10 percent of
                   gross metal payment over $25    1.403
         Treatment charge                          5.403    5.403
         Net value per ton                                 33.627
         52.331 tons at $33.627 per ton                            $1,759.73
         Deductions:               Freight, $4.70 per ton   252.62
                                   Net proceeds            1,507.11
```

Buffalo Hump Group

The Buffalo Hump group of six unpatented claims, owned by W. H. Schweis is in the Desatoya Range about 14 miles by road southeast of Eastgate. It can be reached by automobile over a short road that leaves the Ione Road 10 miles south of the Lincoln Highway. The property was discovered by Thomas Wilson and Robert North some years ago, and although considerable prospecting has been done, the only production has been several carloads of shipping ore. In the first part of 1939 the property was inactive.

Development consists of several adits, the longest 400 feet and subsidiary workings such as drifts crosscuts, and raises, totaling about 1,500 feet. Equipment on the property includes tools for hand-mining, blacksmith shop, and camp accommodations for a crew of several men.

Free gold and a little silver occur in an area said to be about 100 feet wide and 200 feet long in a zone of fractured and altered rhyolite. The rhyolite is considerably kaolinized, but silicification is absent. Insufficient sampling has been done to determine the average gold content or whether the deposit is large enough to warrant operations on a large scale.

The smelter returns on ore shipped from the property by W. H. Schweis to International Smelting & Refining Co. on October 15, 1936, furnished the following data:

```
Metal Quotations       Gold,                       $35.00 per ounce
                       Silver,                     $.4475 per ounce
                                                   Ounces
Settlement assay:      Gold                        0.515
                       Silver                      5.49
                                                   Percent
                       Copper                      .85
                       Lead                        .4
                       Zinc                        Nil
                       Sulfur                      .4
                       Iron                        1.1
                       Insoluble                   92.2
                       Lime                        .5
                       Pounds
Net weight             49,410
Moisture 4.2 percent    2,075
Dry weight             47,335                      23.6675 tons

Metal payment:         Silver, less 1/2 ounce at
                       $0.4475 per ounce[1]        $ 2.233
                       Gold, 91 percent at $35 per ounce  16.403
                       Gross value per ton         18.636
                       Less treatment charge per ton  3.500
                       Net value per ton           15.136

23.6675 tons at $15.136                            $358.23
Deductions:            Freight $3.30 per ton         78.10
                       Net proceeds                 280.13
```

[1]

Building Stone

A deposit of volcanic tuff quarried locally for building stone occurs 3 miles west of the Eastgate ranch and about 1/2 mile off the Lincoln Highway. The deposit is said to be owned by Mrs. Lucille Downing, of Vincennes, Ind. The material was first quarried in the sixties to supply construction stone for buildings at Alpine and Eastgate, and later for several buildings at Fallon.

The quarry has been opened near the crest of a small hill that rises about 100 feet above the surrounding area. The face of the quarry is 100 feet in length and averages 8 feet in height. The tuff is siliceous in character with grain size quite uniform for this class of stone. It is white, weathering to a dull gray, and rather tough, but it is easily cut with an ordinary saw. It hardens a little on exposure, but not enough to raise its compressive strength appreciably.

The rock probably is suitable for building stone when little strength is required, but because of the distance to consuming centers it is of doubtful value in competition with other materials.

FAIRVIEW DISTRICT

The Fairview district is in southeast Churchill County 42 miles by road southeast of Fallon, the nearest railroad point and supply center. The principal mine, the Nevada Hills, is on the west slope of Fairview Peak at an elevation about 5,500 feet above sea level.

The first locations in the district were made by F. O. Norton and associates in 1905. The Nevada Hills mine was located by P. Langsden in January 1906. The discovery of rich silver-bearing float and ore in croppings created considerable excitement, and a boom ensued the following year, which gave the district a temporary population of several thousand. By 1907 the town of Fairview, laid out on the flat west of the mines, had a population of 1,000 and boasted two hotels, several restaurants, stores, and two newspapers. Daily stages and telephone lines connected Fairview with Fallon and Wonder, the latter being 18 miles north. During the first years of the camp's history, mining was in the hands of lessees and numerous wild-cat companies, most of the latter being very short lived. Until 1911, all the ores produced were shipped to smelters for reduction, and in consequence only the higher-grade ores were mined. The freight rate on ore from Fairview to Fallon in 1907 was $12.50 per ton. In 1910 the Nevada Hills Mining Co., incorporated in 1906, acquired control of the Fairview Eagle Mines Co., a continuous property, and the following year the company constructed a 20-stamp mill employing gravity concentration and cyanidation. Electric power was brought into the camp, and water for milling was obtained from wells in Westgate Wash, about 8 miles northeast of the mine. In 1917, after a profitable career, the Nevada Hills property closed because of depletion of the ore reserves. Production of the company from September 1911 to June 1917 is reported to have been $2,265,000. After the property closed, the electric power line and the water line were removed, and the mill was dismantled. Since the Nevada Hills Mining Co. ceased operations, attempts have been made by several companies to revive the mine, the most recent of which has been that of the Nevada Range Mines Co., Inc., the present owners.

In April 1939 activity in the district was confined to small leasing operations in the Nevada Hills property. Most of the patented claims in the district have been taken over by the county for nonpayment of taxes.

Production in the district from 1906 to 1937 was 287,040 tons of ore with a value of $4,171,035, an average of $14.53 per ton. Most of this was produced by the Nevada Hills mine. The annual production of the district is shown in table 3.

I. C. 7093

TABLE 3.— Gold, silver, copper, and lead production from Fairview district, Churchill County, Nev., 1906-37, in terms of recovered metal (Compiled by Charles White Merrill, Mineral Production and Economics Division, Bureau of Mines)

Year	No. of mines	Lode				
		Ore, short tons	Gold		Silver	
			Fine ounces	Value	Fine ounces	Value
1906..	1	479	1,548.00	$32,000	191,045	$128,000
1907..	1	6,543	6,884.37	142,313	640,246	422,562
1908..	4	1,521	2,117.42	43,771	183,592	97,304
1909..	5	1,204	1,706.53	35,277	139,858	72,726
1910..	8	1,153	2,058.71	42,557	160,284	86,553
1911..	6	9,597	3,425.94	70,820	283,411	150,208
1912..	7	30,016	9,694.84	200,410	959,391	590,026
1913..	8	41,952	6,092.37	125,940	674,485	407,389
1914..	8	64,588	5,703.53	117,903	623,832	344,979
1915..	6	65,503	3,981.39	82,303	438,651	222,396
1916..	9	52,688	4,475.23	92,511	356,843	234,803
1917..	4	4,702	806.43	16,670	68,818	56,706
1918..	4	54	38.44	795	2,938	2,938
1919..	2	163	104.53	2,161	6,849	7,671
1920..	1	8	7.49	155	641	699
1921..	3	19	1.31	27	313	313
1922..	3	8	25.54	528	1,621	1,621
1923..	2	107	5.29	109	2,608	2,139
1924..	3	70	6.28	130	1,931	1,294
1925..	5	1,401	120.44	2,490	59,880	41,557
1926..	5	783	93.47	1,932	23,569	14,707
1927..	4	337	35.98	744	6,350	3,601
1928..	2	409	12.72	263	1,988	1,163
1929..	2	975	13.59	281	3,132	1,669
1930..	2	47	3.13	65	1,250	481
1931..	-	-	-	-	-	-
1932..	2	20	26.65	551	533	150
1933..	3	28	27.40	700	587	205
1934..	1	1	2.10	73	1	1
1935..	3	169	97.17	3,401	8,216	5,905
1936..	13	962	511.95	17,918	40,094	31,053
1937..	6	1,533	337.00	11,795	28,949	22,392
Totals	-	287,040	49,965.24	1,046,593	4,911,906	2,953,211

I. C. 7093

TABLE 3.— Gold, silver, copper, and lead production from Fairview district, Churchill County, Nev., 1906-37, in terms of recovered metal (cont'd.)
(Compiled by Charles White Merrill, Mineral Production and Economics Division, Bureau of Mines)

Year	Lode				Total value	Average recoverable value of ore per ton[1]
	Copper		Lead			
	Pounds	Value	Pounds	Value		
1906...	—	—	—	—	$160,000	$334.03
1907...	—	—	—	—	564,875	86.33
1908...	—	—	1,690	$71	141,146	92.80
1909...	—	—	26,602	1,144	109,147	90.65
1910...	2,417	$307	31,602	1,390	130,807	113.45
1911...	—	—	117	5	221,033	23.03
1912...	6,196	1,022	6,538	294	791,752	26.38
1913...	4,292	665	162	7	534,001	12.73
1914...	3,892	518	9,214	359	463,759	7.18
1915...	2,351	411	22,168	1,042	306,152	4.67
1916...	2,407	592	106	7	327,913	6.22
1917...	—	—	10,449	899	74,275	15.80
1918...	99	24	29,932	2,125	5,882	108.93
1919...	—	—	20,780	1,101	10,933	67.07
1920...	—	—	—	—	854	106.75
1921...	—	—	16,558	745	1,085	57.11
1922...	—	—	—	—	2,149	268.63
1923...	174	26	80,962	5,667	7,941	74.21
1924...	129	17	32,659	2,613	4,054	57.91
1925...	1,914	272	911,320	79,285	123,604	88.23
1926...	1,453	203	446,036	35,683	52,525	67.08
1927...	443	58	142,859	9,000	13,403	39.77
1928...	1,112	160	219,156	12,711	14,297	34.96
1929...	498	88	140,207	8,833	10,871	11.15
1930...	55	7	30,348	1,517	2,070	44.04
1931...	—	—	—	—	—	—
1932...	—	—	—	—	701	35.05
1933...	—	—	—	—	905	32.32
1934...	—	—	—	—	74	74.00
1935...	121	10	4,114	165	9,481	56.10
1936...	—	—	3,954	182	49,153	51.09
1937...	—	—	34,000	2,006	36,193	23.61
Totals	27,553	4,380	2,221,533	166,851	4,171,035	14.53

[1] Not to be confused with average assay value of ore.

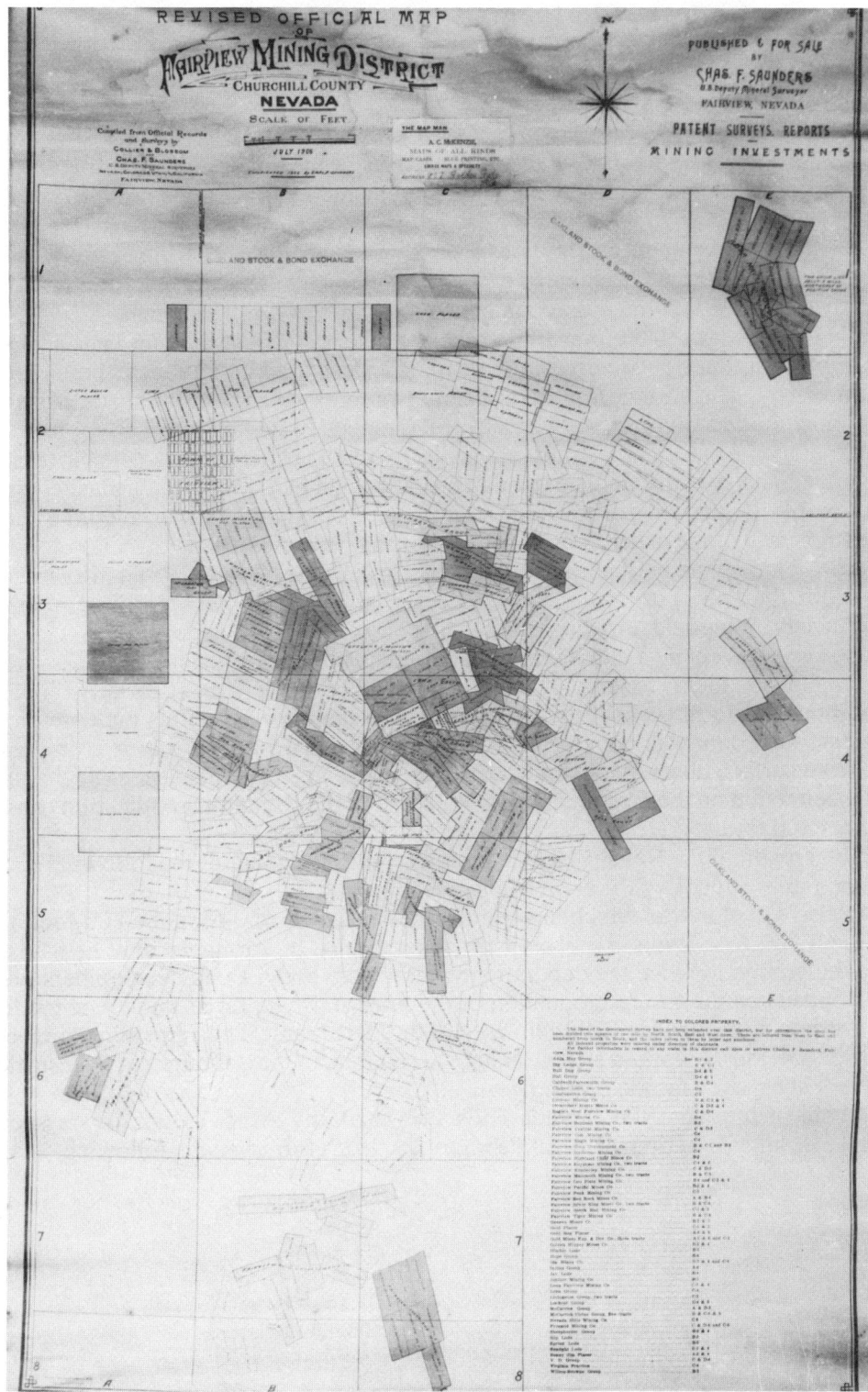

A "revised official map" of the Fairview district published in July 1906 benefitted brokers who sold mining properties. Sales proceeded at a brisk pace if investors could see that their property was located near an already producing ore body in a known claim. Fairview's most important mine—the Nevadahills—is in the exact center of the map; it was the camp's only sustaining producer.

Nevada Range Mines Co., Inc.

The Nevada Range Mines Co., Inc., controlled by Reno interests, comprises 10 patented claims. The property includes the Nevada Hills mine, which has been the principal producer in the district.

The property is developed by shafts to a depth of 1,000 feet. The underground workings total about 9 miles. Equipment includes a Diesel-engine power plant, blacksmith shop, compressor, partly dismantled 25-ton cyanide mill, and a number of camp buildings.

According to Greenan[10], the prevailing rocks are dacite tuff, earlier and later andesite, and rhyolite. Strong fissuring has occurred in the earlier andesite, and along these fissure zones are prominent outcrops. The strongest mineral-bearing fissures strike northwest-southeast and dip south. The most productive vein, the Nevada Hills, ranges in width from 1 to 15 feet. The ore minerals are argentite, stephanite, ruby silver, horn silver, chalcopyrite, galena, tetrahedrite, sphalerite, silver, and gold in a gangue of quartz, calcite, and partly replaced andesite with minor amounts of pyrolusite and rhodochrosite. The average proportion of silver to gold by weight is 100 to 1; as the grade of the ore increases, this proportion decreases.

The Eagle vein, second in importance, roughly parallels the Nevada Hills and averages about 16 feet in width, the richest ore being close to the walls. Other parallel veins of less importance are the Dromedary, Wingfield, and Eagles's Nest.

The vein system is cut by a number of transverse faults having a general northeast-southwest trend and dipping 50° to 75° easterly or westerly. The "Big" fault, a great easterly dipping fault, strikes at right angles to the vein system. Outcrops of the veins west of this fault are prominent, but to the east the country is covered by a later flow of andesite, so that it is impossible to determine from surface observations in what direction the veins are displaced. Segments of the veins have been found on the east or hanging-wall side of the Big fault. Oxidization extends to a depth of about 300 feet.

The annual report of the Nevada Hills Mining Co. for 1915 contains the following statements relative to the future outlook of the company's operations:

The extensive exploration which has been carried on at the 650-foot, the 800-foot, the 900-foot, and the 1,000-foot levels has shown negative results, and roughly may be said to have prospected the ground to twice the depth of any known ore body. Deep development has therefore been stopped and the future production of the mine must, so far as known, come from the shallow workings of the Nevada Hills vein west of the "Big" fault. This remaining ore, occurring in the walls of previously worked stopes, is not measurable, but it is thought to be insufficient to supply the mill at capacity for the coming year.

A shipment of ore from the Nevada Hills vein made by Arnold Dummar, lessee, to the International Smelting & Refining Co. on February 16, 1939, furnished the following data:

Metal quotations:	Silver	$0.64125 per oz.
	Gold	35.per oz.
		Oz. per ton
Settlement Assay:	Gold	0.445
	Silver	37.15
		Percent
	Copper	0.07
	Lead	Trace
	Zinc	Nil
	Insoluble	92.4
	Iron	2.3
	Sulfur	Trace
	Lime	Nil

		Pounds	
Wet weight:		25,100	
Moisture, 2.05 percent		515	
Dry weight		24,585	or 12.293 tons
Metal payment:	Silver, 95 percent		
	@ $0.64125 per ounce	$22.631	
	Gold, 91 percent		
	@ $35 per ounce	14.173	
	Gross value per ton	36.804	
	Treatment charge per ton	4.750	
	Net value per ton	32.054	
12.293 tons @ $32.054			$394.04
Deductions:	Sampling $12.00		
	Hauling 31.38		
Royalty, 15 percent			
of $305.10	45.77		
Freight	45.56		
	134.71		
	Net proceeds	134.71	
		259.33	

Gold Basin Mining Co.

The Gold Basin Mining Co., owned by E. S. Montgomery of Fallon, Nev., consists of five unpatented claims in the Gold Basin section of the Fairview district, several miles east of Fairview Peak and 45 miles southeast of Fallon, the nearest railroad point. The property is accessible over a fair desert road connecting with the Lincoln Highway at Westgate, which is 7 miles S. 20° W. of the mine. The property was discovered about 1924, and although considerable work has been done, the only production has been a few small shipments to the Dayton custom mill at Silver City, Nev.

Development consists of an adit 250 feet long, a winze less than 100 feet deep driven from the adit level, and subsidiary workings comprising in all about 600 feet. There is no equipment on the property except several camp buildings. In April 1939 the property was inactive.

Free gold occurs along a series of fractures having a general strike of N. 30° W. with medium dips N. 60° E. in a quartz latite formation. Along the fractures is a small amount of brecciated and oxidized material that pans well in free gold, probably because of mechanical enrichment from erosion of surface material. The deposits are too small and too far separated to carry the intervening waste rock, so that the outlook for mining any appreciable tonnage of mill ore is not encouraging.

Belle Mountain Mining Co.

The Belle Mountain Mining Co., controlled by W. W. Stockton, comprises the Homestake group of eight unpatented and two patented claims on the north end of Belle Flat, 3-1/2 miles southeast of Fairview Peak and about 9 miles in an airline southwest of Westgate. The road from Westgate to the mine through Gold Basin is impassable, but the property is accessible by car via the Broken Hills road from Westgate, a distance of 30 miles. The altitude is about 6,000 feet.

Although considerable exploring has been done on the claims, there is no record of any production. Workings consist of a main adit with several raises and drifts on the vein, a number of open-cuts, and other workings totaling approximately 1,000 feet. There are three camp buildings on the property, but no mining equipment. In April 1939 the property was inactive.

The vein, consisting of calcite with a little quartz, strikes nearly east and west, with a moderate dip to the south, It is traceable on the surface by open-cuts for a considerable distance and has a maximum width of at least 30 feet. The foot wall is andesite and the hanging wall presumably rhyolite. The economic minerals are gold and silver. The best ore is reported to be along the foothill side of the vein.

Shamrock Group

The Shamrock group of eight unpatented claims, owned by Cyrus Cox of Fallon, Nev., is in a range of low hills 6 miles S. 20° W. from Westgate on the Lincoln Highway and about 2-1/2 miles due east from Fairview Peak. Placer gold was discovered on the claims by the present owner in 1936, and the source of the gold was traced by panning. The only production has been a small amount of free gold that was mortared out of high-grade material by Cox during prospecting.

Development consists of a vertical shaft 64 feet deep, an adit 300 feet long, and minor workings, totaling about 500 feet. Equipment includes a blacksmith shop, a home-made gasoline hoist, tools for hand mining, and accommodations for a crew of several men.

Free gold occurs along fractures in altered rhyolite and rhyolite breccia stained with manganese and iron oxides. The most persistent fractures strike northwesterly and dip nearly vertical. Some of the fractures are accompanied by a gouge ranging from a few inches to over a foot in width.

Westgate Custom Mill

In February 1939 the Westgate Mining & Milling Co., owned by E. S. Montgomery of Fallon and associates, completed the erection of a 35-ton-daily-capacity cyanidation mill at Westgate, situated on the Lincoln Highway 46 miles southeasterly from Fallon, Nev. In April 1939 the mill was operating on custom ores, obtained chiefly from the Nevada Wonder mine with smaller tonnages from the Nevada Hills mine at Fairview, the Gold Ledge mine in the Eastgate district, and other properties within a radius of 50 miles.

The mill is equipped with a 9- by 15-inch Blake-type crusher, a set of 22- by 12-inch Denver rolls, three Snyder disk samplers, a 4- by 4-foot Eimco ball mill, a Simplex classifier, three 10- by 12-foot redwood airlift agitators, four 18- by 10-foot redwood thickeners, a 4- by 7-foot 20-leaf clarifier, a Merrill-Crowe zinc dust-precipitating unit, and auxiliary cyanidation apparatus. Other equipment includes an assay office, a melting furnace, and camp accommodations for a crew of 10 men. Power for milling is supplied by 2 D-11,000 Caterpillar Diesel engines equipped with electric generators. Water for milling is obtained from a well near the millsite.

The custom-milling charge is $4.50 per ton, and payment is based on an average extraction of 90 percent of the gold and 85 percent of the silver contained in the ores.

Manganese Deposit

A bedded deposit of manganese covered by a group of unpatented claims, owned by V. S. Baxter of Fallon, occurs several miles south of the old Fairview townsite, 38 miles southeast from Fallon. It is covered with detritus to a depth of 5 to 30 feet; not enough work has been done to determine its extent. A sample of the manganese analyzed by the Nevada State Bureau of Mines showed 55.2 percent MnO_2, 15.2 percent Fe_2O_3, and 17.9 percent SiO_2.

FIREBALL DISTRICT

The Fireball district is in a low range of unnamed hills in northwest Churchill County about

10 miles northwest of Springer's Hot Springs on the Victory Highway. It was discovered by Fred anderson in 1930. For several years after its discovery a number of properties were prospected by claim owners and lessees for shipping ore, but the results were discouraging and only about 20 tons of ore that averaged $30 per ton is reported to have been produced. In recent years the only activity has been a small amount of desultory prospecting.

The largest property in the district is the Fireball Group, comprising five unpatented mining claims and 200 acres of patented railroad-grant land owned by E. Opdyke of Wadsworth, Nev. Other claims are owned by C. C. Higgins of Reno, Nev.

The economic minerals are gold and a little silver occurring in quartz stringers in silicified limestone cut by rhyolite and andesite.

HOLY CROSS DISTRICT

The Holy Cross district is in southwestern Churchill County near the Mineral County border. Schurz, a station on the Mina-Hazen branch of the Southern Pacific R. R., is 12 miles southwest, and the town of Fallon is 29 miles north. The district was discovered in 1911 by V. B. Terrell, working under grubstake agreement with Craig Catterson, formerly of Rawhide, Nev. Following the discovery considerable prospecting was done, but no important mineral deposits were discovered. The total production, chiefly by lessees, has been about $40,000 in shipping ore. In the fall of 1939 several lessees were active on claims controlled by the Terrell estate.

A diatomaceous earth deposit south of Fallon and several groups of claims northwest of the Holy Cross district proper are included in the Holy Cross district for convenience.

Terrell Group

A group of 26 claims, comprising the principal workings in the district, is controlled by the Terrell estate, administered by Henry Smith of Fallon, Nev.

The workings consist of about 10 shallow shafts and several adits, comprising in all about 3,000 feet of scattered workings. The deepest shaft is 140 feet and the longest adit 1,500 feet. Equipment on the property consists of tools for hand mining and seven camp buildings.

Ore occurs in a series of narrow veins in rhyolite and andesite. The principal metallic minerals are silver chloride, gold, cerussite, a small amount of zinc, and in places some manganese. On the Last Hope claim manganiferous silver veins striking north and south and dipping vertically up to 2 feet in width are traceable on the surface for several thousand feet. The manganese, in the form of psilomelane and pyrolusite, ranges from 5 to 15 percent but the deposits are too small and too low-grade to have economic importance for their manganese content. the silver and gold occur largely in the fines, and lessees obtain a high-grade product by screening and hand sorting.

A small shipment of high-grade screened ore mined from the Silver Star claim by E. P. Frankum and shipped to the Selby smelter of the American Smelting & Refining Co. at Selby, Calif., on August 27, 1938, furnished the following data:

Metal quotations:

	Gold	$35 per ounce
	Silver	.64-1/8 cents per ounce
	Lead	4.90 cents per lb.

		Ounces per ton	
Settlement assay:			
	Gold	1.62	
	Silver	449.92	
		Percent	
	Lead	13.5	
	Insoluble	58.8	
	Iron	5.4	
		Pounds	
Gross weight 12 sacks ore		1,185	
Tare		14	
Wet weight		1,171	
Moisture, 1.5 percent		18	
Dry weight		1,153	
Metal payment:	Gold, @ $31.81663 per ounce		$ 51.54
	Silver, less 5 percent @		
	$0.64 1/8 per ounce		274.09
	Lead, less 10 percent		
	at 3.4 cents per pound		8.26
	Gross value per ton		333.89
	Smelter charge		10.95
	Net value per ton		322.94
1,153 pounds or 0.5765 ton at $322.94		186.17	
Deductions:			
Sampling and assaying		10.00	
Freight		9.32	
Parcel post (sacks)		.40	
		19.72	19.72
Net proceeds			$166.45

Cinnabar Hill Group

The Cinnabar Hill group of six unpatented claims, owned by A. L. Robinson of Fallon, is in a range of low hills 38 miles southeast of Fallon. It is accessible by automobile by driving south from Fallon on the Fallon-Schurz highway to the Camp Terrell turn off, a distance of 25 miles, and thence north- east for about 13 miles, or 6 miles beyond Cushman Well, an old landmark in this section. Cinnabar float was found in this area by Robinson in October 1938; its source was discovered by tracing the float up the side of a small hill. As the property is in the prospect stage of development, there has been no production.

Development work includes a 75-foot shaft, a 60-foot adit, and a small amount of drifting and crosscutting, totaling about 175 feet.

The formation is rhyolite and andesite flanked on the west by granite. The cinnabar occurs in a fissure striking N. 70° E. and dipping about 40° NW. Not enough work had been done to determine the extent of the deposit. The cinnabar occurs in a gangue of quartz, kaolin, and decomposed rhyolite stained with iron oxides.

Bimetal Group

The Bimetal group of three unpatented claims, owned by A. L. Robinson of Fallon, lies 1 mile due south of the Cinnabar Hill property. Gold was discovered here by Robinson in 1932. The only production has been several tons of ore averaging about $40-per ton.

Workings comprise four shafts ranging from 10 to 100 feet in depth and several open-cuts. Equipment includes tools for hand mining and camp accommodations for two men.

Free gold alloyed with a little silver occurs in a quartz vein in granite. The vein strikes S. 70° E. and dips about 55° NE. It is persistent and traceable on the surface for over 1,000 feet, but is narrow, running in width from several inches to a maximum of 2 feet.

Diatomaceous Earth Deposit

A large deposit of diatomaceous earth occurs 17 miles south of Fallon, 1-1/2 miles by road off the Fallon-Schurz Highway. Although the deposit has been known for many years, it has been little prospected, and there has been no production.

The deposit covers a group of several well-rounded hills partly capped by basalt. For the most part, the deposit is covered to a depth of several feet with detritus, consisting of basalt boulders and fragments of-petrified wood mixed with sand. In one place where the diatomaceous earth is exposed on the side of a hill, a thickness of at least 30 feet is indicated. The material is pure white, homogeneous, and apparently of good quality.

I. X. L. DISTRICT

The I. X. L. district is on the east side of the Stillwater Range in central Churchill County. The Silver Hill section on the west slope of the range is sometimes considered a separate district, but in this paper it is included as part of the I.X.L. the I. X. L. district was organized in 1879. The first locations were made in 1878 by Charles S. Kellogg, one of the pioneers of Nevada, who was identified with early-day milling operations at Virginia City.

A number of properties carrying gold, silver-lead, copper, and silver-lead-zinc were worked in this area in former years, but none of them ever passed the prospect stage. A Mexican smelter, one of the first in Nevada, was erected in the district in the early days, but it was unsuccessful. Total production, chiefly from the properties in the vicinity of I. X. L. Canyon, has been about $20,000.

Black Prince Group

The Black Prince group of four patented claims owned by Charles E. Kent of Stillwater, Nev., is at the upper end of I. X. L. Canyon near the crest of the Stillwater Range. The mouth of I. X. L. Canyon is 35 miles north of Frenchman's station on the Lincoln Highway and due west from the Curtis ranch in Dixie Valley. From the mouth of the canyon to the claims, a distance of about 3 miles, the road is impassable by automobile but may be traversed on foot or on horseback. The property was first located by Charles S. Kellogg in 1878, and Kellogg and associates prospected the claims in the eighties. It is reported that some ore was hauled across Dixie Valley to the mill in the Bernice district. The last production was made about 1908, when a small shipment of smelting ore was made. Total production has probably been not more than several hundred tons of ore. For many years the property has been inactive.

The workings consist of the main Black Prince adit about 100 feet in length, several shorter adits, and minor workings, comprising in all about 500 feet of work. The Black Prince adit is open, but the other underground workings are caved and inaccessible. There is no equipment on the property.

The formation is chiefly limestone intruded by granite. The principal showing is in the Black Prince adit, where the mineralization occurs in an irregular contact metamorphic zone, which is traceable on the surface for several hundred yards. The economic minerals are sphalerite, chalcopyrite, galena in a gangue of epidote, garnet, pyrite, quartz, calcite, and abundant magnetite. A 10-pound chip sample cut in the Black Prince adit assayed as follows:

	Ounces
Silver	2.70
Gold	Trace
	Percent
Copper	0.10
Lead	.14
Zinc	2.63
Iron	31.10

A second 10-pound sample taken from an open-cut in mineralized limestone below the Black Prince adit assayed as follows:

	Ounces
Silver	18.94
Gold	.005
Copper	Nil
	Percent
Lead	9.09
Zinc	6.49
Iron	4.76

A large sample of the material on the dump was tested for scheelite under the ultraviolet lamp, but no scheelite was detected.

Bonanza Group

The Bonanza group of 15 claims, owned by Adolph Giannini, is on the north side of I.X.L. Canyon on the east slope of the Stillwater Range. Activity on these claims began in the late seventies, since which time they have been worked intermittently by various owners and lessees with a production of approximately $20,000 in shipping ore.

Development consists of an adit 300 feet long, several shallow shafts, and other widely scattered workings, comprising in all about 1,500 feet. No mining equipment is on the property; in recent years the claims have been idle.

Ore occurs in several quartz and calcite veins in limestone carrying native silver, horn silver, gold, and lead.

Gold Bar Group

The Gold Bar group of four unpatented claims, owned by Charles P. and Leon Cirac of Fallon, is in Cox's Canyon about 1 mile south of the Revenue group of fluorspar claims. The Cirac Bros. have owned the property for 30 years, and except for a small amount of gold obtained by mortaring and hand-panning, there has been no production.

The workings consist of several adits, the longest about 100 feet, and some scattered prospect holes totaling not more than 200 feet. Equipment on the property consists of tools for hand mining and camp accommodations for two men.

The formation is shale and slate intruded by granodiorite. The sediments are locally crumpled and folded, and in places the shale contains numerous cubes of hematite, pseudomorphic after pyrite. Several quartz veins, up to a maximum of 30 feet in width, with variable strike and dip, occur within the sediments. Insufficient sampling has been done to determine whether the vein material warrants large-scale operation. The economic mineral is gold which occurs in a free state.

Revenue Group

The Revenue Group of four unpatented fluorspar claims is on the north side of Cox's Canyon on the west slope of the Stillwater Range, 23 miles northeast of Stillwater. the deposit was discovered late in 1938 by Cirac Bros. In April 1939 they were prospecting the claims with the object of producing a shipping product by hand sorting.

At the time of the writer's visit, the workings comprised several trenches and an adit 10 feet in length, which was being driven under the surface showings.

The prevailing formation is shale and limestone intruded by a fine-grained basis ditch, which has an undulating outcrop traceable for about 1,000 feet; in places, fluorspar up to 5 feet in width occurs along both sides of the dike. Near the surface, the spar is considerably mixed with detrital material, but boulders of solid spar up to 60 pounds in weight had been excavated from the shallow workings. In places the limestone adjacent to the dike is traversed with a network of fluorspar veinlets over widths up to 20 feet, but this material would have to be concentrated to make a commercial product. The fluorite is green, purple, white and black. Although only a small amount of work had been done, the showings appear very promising, and when a greater depth is obtained it is probable that a commercial product can be obtained by hand-sorting. Working of the deposit is handicapped by poor transportation facilities, since the road from Stillwater is in poor condition and impassable in wet weather. The most convenient shipping point is Fallon, on the Southern Pacific R. R., 38 miles southwest.

JESSUP DISTRICT

The Jessup district is in the range of low hills at the southwest end of the Trinity Range in northwestern Churchill County. Huxley station, on the Southern Pacific R. R., is 10 miles southwest. The district is easily accessible by automobile over a desert road 4 miles in length, which branches off the Victory Highway at a point 26 miles southwest of Lovelock, the county seat of Pershing County.

The first location in the Jessup district was made by Frank Jessup and L. H. Murray in 1908, and in the same year John Macedon and associates shipped several carloads of ore from the Gold King claim reported to have averaged better than $100 per ton. Considerable activity in the district followed, and a number of small companies were organized, which, with lessees, were active in the camp in 1908 and 1909. Although considerable shallow development work was done over an area at least 1 mile in length and 1/2 mile in width, the results were not encouraging, and the camp became inactive except for sporadic leasing and prospecting operations. The total production of shipping ore from the district, largely from the Gold King claim, has been about $15,000.

Groups of claims are held in the district by Charles Polk, Albert Loose, H. O. Westergard, Emil Stanc, Geo. W. Lang, Olaf Johnson, and others from Lovelock, Nev. The claims have been prospected by at least 20 shallow shafts and a number of adits and open-cuts totaling about 3,000 feet. None of the workings are more than 200 feet below the surface. No water is available in the immediate vicinity of the camp and must be hauled for domestic use from Hot Springs station on the Victory Highway or other places. In the fall of 1938 the only activity in the area was on the Valley King group of claims, which was being worked by lessees.

Valley King Group

The Valley King group, consisting of five unpatented claims, is owned jointly by George W. Lang and Olaf Johnson of Lovelock. The Gold King, the original discovery claim adjoining the

Valley King group, is owned by George W. Lang individually. In 1938 the Valley King group was under lease to Kenneth Dale and Dick Collins. The lessees, after mining and shipping about 30 tons of ore, averaging $40 per ton, to the Dayton custom mill at Silver City, Nev., relinquished the lease.

The Valley King single compartment shaft attained a depth of 150 feet. Other scattered shafts and subsidiary markings total about 1,000 feet. No equipment other than tools for hand mining is on the property.

The formation is chiefly andesite and rhyolite. The economic minerals are gold and silver occurring in small veins ranging in width from a few inches to 1 foot. The principal vein strikes N. 25° E. and dips 60° to 70° SE. The gold is in a free state, and the silver occurs as cerargyrite in a gangue of quartz and crushed country rock impregnated with iron oxides. A small amount of scheelite occurs in the vein material.

In 1932 Olaf Johnson discovered placer gold on the top of a small hill on the Valley King claim, The alluvium, largely clay with a few small boulders and well-rounded pebbles, is cemented with lime. Several ounces of gold have been produced by dry-washing methods. The gold which is associated with black sands, has a fineness of about 600.

Diatomaceous Earth

Diatomaceous earth outcrops along the eastern foothills of the Trinity Range north and south of Jessup for 15 miles. the deposits appear to be very extensive laterally, but little work has been done to determine their thickness. For the most part the diatomaceous earth is covered with alluvial material eroded from the nearby mountains; where exposed, it is mixed with some volcanic ash. The only production has been a few carloads used locally for insulation purposes.

LAKE DISTRICT

The Lake district is on the east side of the Humboldt Sink at the southern extremity of the Humboldt Range. The salt-producing section near Huxley station on the Southern Pacific R. R. is known as White Plains Flat.

The principal production from the district has been salt from salines in the Humboldt Sink. A small amount of shell limestone also has been produced for agricultural purposes from deposits near Huxley station on the Southern Pacific R. R. Sodium nitrate also occurs in the district, but none has been produced.

Limestone Deposit

A deposit of shell limestone occurs on the southwest flank of the Humboldt Range about 1 mile northeast of Huxley station on the Southern Pacific R. R. The main part of the deposit probably is on patented railroad land. A small tonnage of limestone for agricultural purposes was mined about 30 years ago and shipped to the Pacific coast. The remains of an abandoned kiln on the deposit is evidence that an attempt was made to burn the limestone locally for lime.

The deposit is of lacustrine origin. composed of fresh-water mollusks. Where exposed, it covers an area approximately 3,000 feet long and 1,000 feet wide. The deposit was formed in an arm of ancient Lake Lahontan and,judging from its origin, is probably not more than 10 feet thick. A number of scattered open-cuts expose the material to a depth of 4 to 6 feet. Because of its proximity to the railroad, its purity, and the ease with which it can be mined, the deposit is of potential value as a source of limestone for agricultural use on the Pacific coast.

White Plains Salt Deposit

The White Plains salt deposit, on the main line of the Southern Pacific R. R., has been a notable source of salt production in Nevada. It was discovered by Walter Schmidt in 1870, and shortly after it was acquired by the Desert Crystal Salt Co., which operated for many years. The salt was produced by solar evaporation in a series of vats dug in the surface of Humboldt Sink. the vats comprised a total length of 8,500 feet and were 55 feet in width. The brine was pumped into the vats from salt springs in the vicinity. Beginning in 1911, the International Salt Co., operating under lease from the Desert Crystal Salt Co., produced small amounts of salt annually for several years; the last production of record was in 1915. Although the bulk of the salt obtained here was used in the reduction of silver ores in the early days, some salt was also refined for domestic and dairy purposes.

In addition to the springs, salt in the form of ar incrustation on the surface of Humboldt Sink covers an extensive area. Although the reserves are no doubt large, the deposit is too remote from large consuming centers to compete with other sources of supply.

Sodium Nitrate Deposit

Sodium nitrate has been found over a considerable area on the southeast side of the Humboldt Sink in the low hills constituting part of Humboldt Range. One of the hills, of a marked red color, is known as Niter Butte. The existence of nitrate salt in this locality is said to have been known to the early Indians, who used it for its supposed medicinal value. Walter Schmidt located the first nitrate claims in the district in 1868. Attention was called to the nitrate deposits in the eighties by an encouraging report prepared by B. B. Redding, which was read before the California Academy of Sciences. A company known as the Nevada Niter Co. was organized and acquired an extensive acreage of nitrate lands, but after a short period of desultory prospecting the venture was abandoned. Interest was revived about 1902, when a group of Lovelock residents located and patented 320 acres.

This ground was bonded to Detroit interests, which organized the American Nitrate & Potash Co. This company did a small amount of work, but as results proved disappointing the project was abandoned. There is no record of any production of nitrates, and for many years there has been no activity on the ground. A description of the deposits is given by Gale.[11] The sodium nitrate occurs in small irregular seams and bunches in places that are protected from weathering. It is associated with other salts, chiefly sodium chloride. The occurrences of the nitrate are small and scattered, and the outlook for their commercial exploitation is discouraging.

LA PLATA DISTRICT

La Plata, also known as the Mountain Well district, is in central Churchill County on the eastern slope of Stillwater Range, about 12 miles in an air line a little west of north from Bermond (Frenchman's station) on the Lincoln Highway (U. S. Route 50). The nearest railroad station is Fallon, about 50 miles west by road via Dixie Valley. The old road from Fallon via Stillwater is considerably shorter, but part of the road from Stillwater to Mountain Well near the summit of the range is in poor condition and seldom used.

The La Plata district, discovered in 1862, attained considerable prominence as a boom camp during the middle sixties, but there is little evidence to show that any appreciable amount of ore was produced. The town of La Plata, established about 1863, was the county seat of Churchill County from 1864 to 1868. In 1863 and several years afterward many claims were located, many of which were sold to eastern capitalists, who did considerable prospecting, but the general results were discouraging. The county seat was removed to Stillwater in 1868, and the following year most of the miners deserted the district for the White Pine boom in the eastern part of the State.

In 1864 the Silver Wave Mining Co. erected a 10-stamp mill at La Plata at a cost of $150,000, but little evidence, such as tailings or extensive mine workings exists to indicate a large production. This mill was removed subsequently to the Ellsworth district in Nye County. Another mill was built in Eleven-Mile Canyon, several miles north of La Plata Canyon, about 1864, which likewise was unsuccessful, presumably for lack of ore.

From 1869 to the present, the only mining activity has been sporadic prospecting operations. Over the last 35 years mineral production statistics show several small shipments of silver ore. In April 1939 the district was inactive. Fluorite was discovered there in 1939.

La Plata Mine

A group of unpatented claims, comprising the old La Plata mine and mill-site, was relocated in January 1939 by F. J. Sur, Lewis Barr, and J. C. Stewart of Fallon, Nev.

Development consists of an incline shaft, several adits, and other workings, totaling approximately several thousand feet. Some of the workings are caved and inaccessible.

The prevailing formation is shale intruded by granite. The La Plata vein, as exposed in a small stope open at the surface, is in granite. The vein strikes E. and W. and dips 60° N. Other quartz veins, either in the granite or in the shale outcrop in the vicinity, have been prospected superficially by open-cuts. The economic mineral is silver.

Michigan Claim

The Michigan claim was located by Mark Harris and Fred T. Pine, of Westgate, Nev., in the early part of 1939 as a fluorspar prospect. The claim is in the lower part of La Plata Canyon about 2 miles below the old La Plata townsite. Only location work had been done on the claim to April 1939.

Limestone intruded by rhyolite is the prevailing formation. Fluorspar occurs in a series of small fractures in the limestone striking about east-west and traceable on the surface for several hundred feet. From surface showings, the fluorite veins are too small to be commercially important.

LEETE DISTRICT

The Leete district is in northwest Churchill County at Leeteville on the Southern Pacific R. R. 18 miles east of Wadsworth in Washoe County. In former years the Playa lake from which salt was produced was known as the Eagle salt marsh, while that which produced borax was called Hot Springs borax marsh. Salt was first produced here in 1870 by a man named Leete, who discovered the deposit and organized the Eagle salt works. Large quantities of salt were produced annually for many years to supply the silver mills on the Comstock and later for dairy and domestic use. This deposit, because of better transportation facilities, supplied salt to the Comstock mills in competition with other sources of supply, particularly Sand Springs Marsh. (The Central Pacific, now the Southern Pacific R. R., was completed May 10, 1869.) The Eagle salt works continued to produce salt intermittently until 1915, when operations ceased. A small amount of borax was produced from the salines in 1871, but the venture was unsuccessful. From 1879 to 1884 the production of salt amounted to 334,000 tons, and the total production probably has exceeded 500,000 tons.

The playa lake in which the salt and borax occur is a desert mud plain of recent age covered during the summer months with a white saline efflorescence. The salt was recovered by solar evaporation. The saturated brine occurring about 20 feet below the surface was pumped into shallow excavated vats enclosed by low embankments. The vats were 50 feet wide and about 100 feet long and covered a considerable area. The saturated salt solution crystallized on the sides and bottoms of the vats, and, by successive evaporations, a layer of salt was formed that was thick enough to be hoed into piles and shipped without further refining. One acre of vats produced 10 tons of salt daily during good weather. The season lasted from June to October.

SAND SPRINGS DISTRICT

The Sand Springs district is 25 to 30 miles southeast of Fallon, the nearest railroad point. The Sand Springs salt marsh in this area is a playa lake covering nearly 40 square miles in Salt Springs Valley, known in the early days as Alkali Valley and recently as Four-Mile and Eight-Mile Flats. Salt production began in 1863; previous to 1870 considerable quantities of salt for the reduction of silver ores on the Comstock were produced. After 1870, salt for metallurgical purposes could be obtained more cheaply from the deposits in the Leete district because of better transportation facilities; therefore salt production at Sand Springs was discontinued temporarily. In recent years the Sand Springs salt field has been worked by small companies and individuals to supply salt for local dairy and table purposes.

Borax was discovered in the Sand Springs salt marsh about 1869, and the American Borax Co. erected a plant, which operated for several years.

The gold deposits of the Dan Tucker property and vicinity were first prospected in 1905 by C. W. Kinney. The production of metals, chiefly gold, has been about $30,000, most of which was from the Dan Tucker mine.

Dan Tucker Mine

The Dan Tucker mine comprises a group of 5 unpatented claims situated near the Lincoln Highway 31 miles southeast of Fallon, Nev., the nearest railroad station and supply center. Although the property was prospected by C. W. Kinney in 1905, very little work was done until 1912, when Leslie L. Leonard and C. W. Kinney sank a 100-foot shaft. The first production was made in 1919 by lessees, who shipped three carloads of ore yielding $215 to $300 per ton. The Dan Tucker Mining Co. was organized in 1925, and in the following year it leased the mine to Smith, Towle, and Young, who in 1927-erected a small amalgamation mill at Sand Springs in which 1,000 tons of ore was treated. In 1931 the property was acquired by another company, and in 1938 it was awarded to E. E. Tailleur, Fred Tailleur, and Dick Kemp on a labor lien. The owners proceeded to work the mine and shipped 10 carloads of rich ore. In the fall of the same year those connected with the Bralorne Mines, Ltd., of British Columbia, obtained a lease and bond and organized the Summit King Mines, Ltd., a subsidiary, to operate the Dan Tucker and other properties in the vicinity. In April 1939 the company had 12 men on development. Production from the Dan Tucker mine is reported to have been about $30,000, mostly shipping ore.

Workings comprise three shafts, the deepest of which is 200 feet, and subsidiary workings totaling approximately 1,000 feet. Equipment includes a gasoline hoist, a two-stage, air-cooled compressor, a blacksmith shop, and several camp buildings.

The formation is schist, limestone, and andesite. Free gold occurs in veins in a silicified zone striking east and west with a moderate dip to the south. The zone is traceable on the surface for several miles. The economic minerals are silver chloride and gold in a gangue of sugary quartz and crushed andesite.

Salt Deposits

The following data on the production of salt from the sand Springs marsh in the early days are from Browne.[12]

The most productive salt bed at the present time is that of the Sand Spring Salt Mining Co., 75 miles east of Virginia City. The claim-of this company, consisting of 1,600 acres, occupies a depression in the southeastern corner of an extensive alkali flat, the center of which for a space of several hundred acres is damp and marshy, and some portion of it is covered with a few inches of water. This damp surface is coated to a depth of two or three inches with a crystallized incrustation of salt with which the clayey strata below are changed. On removing this coating of salt, a thin body of fine white clay is exposed overlying a stratum of soft black clay, which in turn rests upon a seam of green and black clay containing coarse globules of salt. Beneath this seam occurs a deposit of crystallized salt, hard and massive but of unascertained thickness. In collecting the salt at this place, a tract embracing a score of acres or more is selected and divided into strips, from each of which, in regular order, it is scraped into large heaps with a broad wooden hoe. These heaps, after being exposed for a few days to drain and dry, are conveyed in wheelbarrows or cars running on wooden tracks laid down for the purpose of facilitating transportation over the soft ground, and thrown upon platforms or dumps, when the salt is ready for sacking and shipment to market. after one of the surface sections has been stripped of salt, the incrustation immediately begins to reform, and so rapidly do the secretions from the saliniferous clays below proceed, that a few weeks, and sometimes less, is sufficient to fully replace it, admitting the gathering of a fresh crop at least every month. As the masses of crystallized salt underlying these clayey strata are probably inexhaustible, it would seem as if this process of replenishment might go on forever.

Prior to 1862, all the salt used in Nevada was brought from San Francisco, at an average cost, laid down in Virginia City, of about $150 per ton. During that year parties having imported a herd of camels or the purpose began packing this commodity in from salt pools, 45 miles southeast of Walker Lake (Rhodes marsh in Mineral County), whereby the price was somewhat reduced. The following year, the Sand Springs Salt Co. having commenced operations, the price of salt suffered a further reduction, and for the past two years salt has been delivered to the mills about Virginia City at the uniform rate of $60 per ton, being considerably less than the average cost of freight from San Francisco. During the year 1865, this company disposed of about 150, and during the past year 250 tons of salt per month, most of which was consumed in the mills and reduction works, a little, also, after grinding, having been used for meat packing and culinary purposes, for which it is well adapted. The Sand Springs Salt Co. has over $100,000 invested-in-this business, and though owning several large teams, hire many others to haul the product of their salt fields to market, their freight bills amounting to from $10,000 to $15,000 per week. Large as is the amount of salt they are thus enabled to deliver, the supply is scarcely equal to the demand, some of the larger mills consuming between 35 and 40 tons per month.

Sand Springs salt marsh is a playa lake formed by the evaporation of a shallow body of water in an enclosed basin. During the winter months water collects on the surface of the basin, forming a sheet of brine perhaps 15 or 20 square miles in area, but during the summer this water evaporates, leaving a layer of salt several inches thick. The salines consist largely of sodium chloride, sodium sulfate, and sodium carbonate, with smaller amounts of sodium borate, potassium chloride, and calcium compounds. Magnesium salts are characteristically absent. This playa, as well as others in the Great Basin region, originated by repeated desiccations rather than by the sinnabar evaporation of the deep lake, and in consequence the salts are considerably intermixed with clay and silt brought into the basin by surface drainage of the surrounding areas.

In recent years smaller quantities of salt have been collected annually from the surface of the marsh to supply the local demand for stock and dairy purposes. It is gathered by shoveling into trucks and does not undergo any refining process.

Borax Deposits

The recovery of borax from Sand Springs marsh was one of the pioneer projects of this nature in the Great Basin region. The borax plants at Sand Springs operated in 1870 and 1871 with a capacity of about 1 ton of borax per day. Production ceased in 1872 because of a drop in price of boracic acid from 30 cents to 9 cents per pound.

The most characteristic borate mineral of the playa-lake deposits, such as Sand Springs marsh, is ulexite, a hydrous borate of sodium and calcium, commonly referred to as cotton balls, occurring around the periphery of the marsh near the surface. The material worked in the early days contained about 10 percent borax, although some of it yielded as much as 30 percent.

The process for the recovery of borates from the saline deposits was simple; the cotton balls were either picked by hand from the shallow excavations in the mud, or the richest portion of the boraciferous mud was shoveled from the surface, and the material thus collected was hauled to semicircular iron pans set on a foundation of brick and fired from beneath with sagebrush or pinion. The pans were charged with water and crude borates and stirred with poles until the soluble salts dissolved, after which the liquor was drawn off into crystallizing vats wherein the borax was crystallized out on wires. In deposits where the soluble carbonates were not present in excess, sodium carbonate was added in the firing pans to break up the lime and boric acid as found in ulexite. The borax obtained by crystallization was again refined by a second crystallization in the same manner.

SHADY RUN DISTRICT

The Shady Run district is on the western slope of the Stillwater Range 30 miles northeast of Stillwater and 40 miles southeast of Lovelock, the nearest railroad connection. It embraces an area lying between Fondaway Canyon on the north and Mill Canyon on the south. The I.X.L. district adjoins it on the south and the White Cloud district on the north. Although a number of gold and silver-lead prospects were worked in this area in the early days, no deposits of commercial importance were found and there is no record of any production. A small custom mill was erected in Mill Canyon in the 1880's by a group of miners from Virginia City, including John Fondaway, G. W. Humphrey, and Isaac Zintmeyer, but it was unsuccessful. Probably the last work in the district was done about 1916 by D. G. Zinn and associates on a gold prospect on the north side of Fondaway Canyon. Here the vein is in quartzite near a quartz-porphyry intrusive. In Shady Run Canyon, south of Fondaway Canyon, are several gold prospects in quartzite. According to John T. Reid of Lovelock, a short distance south of the mouth of Fondaway Canyon near the summit of a small butte, isolated from the main range, is a flat vein that in former years was prospected for gold by K. B. Jenkins and associates.

SODA LAKES DISTRICT

The Soda Lakes are on the edge of the Carson Sink 2 miles northeast of the former site of Ragtown—an emigrant station on the Overland Route to California—7 miles northwest of Fallon. The occurrence of soda in Little Soda Lake was discovered by Asa. L. Kenyon in 1855, and in the late sixties borax was discovered in the brine of Big Soda Lake by William Troup. In 1868 Kenyon sold Little Soda Lake to a San Francisco group, which organized a company and began the production of soda. This was the first production of soda in Nevada and probably the first of any importance in the West. The production of soda from Big Soda Lake was begun in 1875 and continued until 1893. The combined annual yield from both places ranged between 300 and 800 tons of soda between 1868 and 1893. About 1869 William Troup and associates attempted to produce borax from Big Soda Lake, but this venture was unsuccessful.

According to Russell,[13] the two lakes occupy deep depressions formed by extinct volcanic craters. Big Soda Lake is nearly circular, with an area of 268-1/2 acres; Little Soda Lake is smaller and varies considerably in size according to weather conditions. The rim of the larger lake rises 80 feet above the surrounding desert and was 165 feet higher than the surface of the lake when the latter was 147 feet deep. The volcanic cones of basalt were not formed during a single eruption but have a long and complicated history; they are perhaps sublacustrine in their origin. There are no streams either tributary to or draining these lakes; the water in them is supplied almost entirely from subterranean sources. The rocks, pebbles, and organic matter along the shore lines of the lakes are coated white with crystals of gaylus-site ($Na_2CO_3CaCO_3\ 5H_2O$) and trona ($HNa_3(CO_3)2.2H_2O$). At the time of Russel's investigation the larger lake was calculated to contain about 2,000,000 tons of salts, of which 428,000 tons was sodium carbonate, 342,000 tons sodium sulfate, and 1,284,000 tons sodium chloride, these being the principal salts of economic importance. The brines of Soda Lakes have been considerably diluted in recent years by infiltrating waters from the Truckee-Carson irrigation project.

The method to produce sodium carbonate from Soda Lakes has been described by Knapp[14] and, briefly, was as follows:

The brine carrying the sodium carbonate and other salts in solution was pumped into vats dug in the clay beds in the vicinity, where it was evaporated by solar heat. After the solution had attained a density of 29° to 30° F., it was conducted to a crystallizing vat. If the solar heat was sufficient to keep the temperature up to 75° F. or above, the density was run up still higher. As crystallization of the sodium carbonate progressed, strong solution was added at intervals and the waste solution was decanted. During warm weather a hard cake of the sesquicarbonate of soda accumulated to a depth of a few inches to 1 foot in the bottom of the crystallizing vat. With the approach of cold weather the cake was broken, taken out of the vats, washed, and dried in the air. The air-dried material had the following composition:

	Percent
Bicarbonate of soda	36.46
Carbonate of soda	45.96
Chloride of soda	.32
Sulfate of soda	1.25
Water	16.16
Insoluble	.02
Total	100.17

After the material was air-dried, it was heated in either a revolving or a reverberatory-type furnace to drive off the water of crystallization and excess carbonic acid, leaving a nearly pure anhydrous monocarbonate of soda, or soda ash, which was ground and sacked for shipment.

The foregoing process is based on the principle that if bicarbonate and carbonate of soda are present in a dense solution, 1 molecule of bicarbonate will unite with 1 molecule of carbonate to form sesqui carbonate, or trona, which precipitates, while sodium chloride will remain in solution unless the solution density rises above 32° F., and sodium sulfate will remain in solution up to 34° B., providing the temperature stays above 75° F.

By the foregoing process only about 25 percent of the available soda is recovered, but since the cost of handling the solution was merely a question of pumping, the low extraction was not a cause for great concern.

Some soda was also obtained in solid form from Little Soda Lake, where it occurred in a layer about 2 feet thick near the surface.

TABLE MOUNTAIN DISTRICT

The Table Mountain, Bolivia or Boyer district is in the Stillwater Range in north-central Churchill County near the Pershing County border. The central part of the district is 57 miles north of Bermond (Frenchman's station) on the Lincoln Highway. It is also accessible from Lovelock on the Victory Highway or from Winnemucca on the Southern Pacific R. R. The distance from Winnemucca to the Boyer ranch in Dixie Valley is 87 miles; the road passes through Grass and Pleasant Valleys.

The district covers a large, vaguely defined area that contains a variety of minerals, including nickel-cobalt, copper, lead-silver, antimony, titanium, gold, and kaolin. From the viewpoint of past production, the gold deposits have been the most important.

Nickel-Cobalt Deposits

There are two nickel-cobalt occurrences in the Table Mountain district that were of sufficient importance to induce a number of attempts at exploitation in former years. The deposits are situated in the vicinity of the old camp of Bolivia about 3 miles up Cottonwood Canyon on the east slope of the Stillwater Range. The mouth of Cottonwood Canyon is 1-1/2 miles N. 35° W. from the Boyer ranch in Dixie Valley. There is a uniform grade up the canyon, but in several places the road has been washed out and some repair work is necessary to make it passable for automobile. Cottonwood Creek of variable flow, is fed by melting snows and springs; the flow may exceed a thousand gallons per minute during certain periods.

The nickel mine at Camp Bolivia was discovered in 1880 by John Mason, Charles Bell, and his brother, William. The Lovelock or Cobalt mine, about 1 mile west, was located at the same time by George Lovelock, Sr. The mines were prospected for a number of years, following their discovery, and the first ore is said to have been shipped to Swansea, Wales, from the Lovelock mine by W. S. Keyes, a prominent early-day mine operator from Eureka, Nev. In 1887 an English company bonded the property and shipped a few tons of rich ore but did not complete the purchase. Subsequently,, it was again bonded to an English syndicate, which organized the Nevada Nickel Co., which shipped about 15 tons of ore to San Francisco. In the late eighties the company erected a sulfuric acid leaching plant at the mine at a cost of $50,000, but this venture was unsuccessful owing to the fact that the Italian chemist who designed the plant was unfamiliar with the metallurgy of nickel-cobalt ores. Later, a 5-ton-capacity water-jacketed furnace was erected at the Nickel mine, but it blew up a short time after it was placed in operation. In the late eighties a small smelting furnace was also erected at the Lovelock mine by George and Frank Bothwell and associates from New York, and from the size of the slag dumps not more than 100 tons of ore was treated. The Lovelock mine was last worked by the Mines Development Go. of Nevada. This company produced several carloads of ore that assayed as high as 26 percent copper with a small amount of nickel and cobalt. From the same vein that produced the copper ore, a few tons of nickel-cobalt ore was mined that yielded 29 percent nickel. It is also reported that 500 tons of nickel-cobalt ore was shipped from the Lovelock mine to Swansea, Wales, in the early eighties.

The mines were worked fairly continuously from 1880 to 1890, when they closed because of litigation and reopened in 1904 only to close again in 1908, since which time they have been inactive. Several years ago the Lovelock mine, comprising two patented claims, and the Nickel mine, of four unpatented claims, were purchased from the county for delinquent taxes by Tasker L. Oddie and Fred H. Luetjens of Reno, Nev.

The workings at the Nickel mine comprise about six short adits and an inclined shaft about 100 feet deep. None of the workings are more than 100 feet from the surface. There is no equipment except several dilapidated camp buildings and the ruins of the old leaching plant, consisting of mill building, leaching vats, and boiler.

In the vicinity of the Nickel mine the formation is diorite, andesite, and fine-grain quartzite, the latter forming a bold and jagged ridge above the mine workings. Some quartzite on the dumps shows the characteristic stain of arsenic minerals, with little evidence of cobalt-nickel mineralization. The nickel-cobalt ore occurs in small stringers up to several inches in width in the sheared and brecciated andesite near a fault contact striking N. 45° E. and dipping NW., the diorite forming the footwall. According to analyses made by Professor Newberry[15], the samples from the greatest depth contain niccolite, which in the upper levels consists entirely of hydrated arseniate, or annabergite, containing 33.71 percent nickel oxide, 36.44 percent arsenic acid, and 24.77 percent water. The sample lot of 15 tons mined in the early days contained 12 percent nickel, 7 percent cobalt, and 29 percent arsenic. the material on the dumps is all oxidized; no sulfide minerals were detected. A 10-pound sample gathered from the dumps by the writer assayed 7.13 percent nickel and 0.33 percent cobalt.

At the Lovelock mine the ore is more complex and in addition to nickel and cobalt contains copper. A number of specimens of erythrite (cobalt bloom) were found on the dumps. Here, likewise, the mineralization occurs in the andesite near a contact with diorite striking N. 40° E. and dipping 50° to 60° NW. A number of ramifying seams of gypsum occur in the andesite. Two 10-pound samples taken from the mine dumps contained 0.40 percent nickel and 0.25 percent cobalt, and 0.46 percent nickel and 0.26 percent cobalt, respectively

Workings of the Lovelock claims are all superficial in character and total approximately 1,000 feet.

Copper Deposits

A number of copper prospects occur on Treasure Box Hill at the head of Bell Mare Canyon south of Cottonwood Canyon. Deposits are reached on horseback up Cottonwood Canyon past the nickel properties. From Boyer ranch the distance is about 15 miles. The principal properties were located in the early sixties by Alva Boyer, C. S. Kellogg, Jacob Strananger, and Patrick Reid. In the early days several wagon trains of rich, hand-sorted copper ore were mined from surface workings and taken to Sacramento for transportation to Swansea, Wales. In about 1900, a group from Colorado erected a small smelting furnace on the Azurite-Nevada Queen group of claims, which is still intact, but, judging from the condition of the smelter, no ore was reduced. About 1910 the Boyer Copper Mines Co. acquired control of 49 claims and did some development work and diamond drilling, but there was no production. In recent years claims have been idle.

The Treasure Box group of six patented claims is owned by the V. A. Twegg estate, the Crawfton Uniace estate, and John T. Reid, the latter of Lovelock, Nev. The adjoining Azurite-

Nevada Queen group of five patented claims is owned by Tasker L. Oddie and Fred H. Luetjens, both of Reno, Nev.

The Azurite-Nevada Queen claims have been prospected by several adits and a number of open-cuts and the Treasure Box group by an old shaft reported to be 170 feet deep, two adits, one 800 feet long, and subsidiary drifts, raises and winzes, totaling on the two groups of claims about 1,500 feet of workings. Most of the working are caved and inaccessible. There is no equipment on either property except the old smelter.

According to Carpenter,[16] the copper ore occurs near the contact of andesite porphyry flow, overlying an earlier mass of strongly faulted green andesite. The main work on Treasure Box Hill was done in a bed of copper-bearing andesite about 100 feet thick. The contact dips about 20° NW. The ore is chiefly chalcopyrite disseminated through the green andesite and, according to Carpenter, the lower 30 feet of the bed averages nearly 5 percent copper and $1 in gold with a trace of silver. In places just above the contact small, iron-capped veins occur in the green andesite and the andesite porphyry. When followed downward, these veins lead to massive chalcocite disseminated in a gangue of breccia, and pieces of pure black sulfides as large as a man's hand have been found occasionally. Massive bornite, tenorite, and cuprite also occur mixed with the two carbonates—malachite and azurite,—making small deposits of rich ore. Development by the Boyer Copper Mines Co. exposed a block of ore 200 feet long, 100 feet wide, and 500 feet deep, measured on the dip, containing 1.7 percent copper and about $0.70 in gold per ton.

Dixie Comstock Mine

The Dixie Comstock mine, comprising 10 claims, is on the west side of Dixie Valley in the foothills of the Stillwater Range, about 45 miles north of Bermond on the Lincoln Highway. The property was discovered in April 1934 by Clyde Garrett, and shortly thereafter the controlling interest in the property was acquired by the Comstock Keystone Mining Co. of Virginia City, Nev. In the spring of 1935, a 30-ton-daily-capacity amalgamation mill was erected at the mine, and late in the summer of the same year flotation equipment was added. Incomplete data indicate that the production of gold and a little silver had been about $150,000 in shipping and milling ore. In 1939 the tailings pile, containing about 6,000 tons, was being reworked by the company, and several lessees were employed in the mine.

Development includes a 200-foot incline, a vertical shaft about 100 feet deep, and drifts, winzes, and other workings, totaling about 1,500 feet. Mining equipment includes a 12-by 10-inch single-stage compressor, a 17-horsepower Hercules geared hoist, and a blacksmith shop. Milling equipment consists of a homemade ball mill 7 by 5 feet, 9-by 12-inch Blake-type crusher, simplex classifier, 4-by 10-foot, amalgamating plate, and four flotation cells. Power is furnished by an 80-horsepower 2-cylinder Fairbanks-Morse Diesel engine belt- connected to a 30-kilowatt alternating-current generator. Camp buildings at the mine can accommodate a crew of 15 men. Water for milling is supplied by a well 30 feet deep near the mine.

Gold, associated with a little silver, occurs in a large vein in a hydrothermally altered igneous formation. Mining is hindered by the intense heat and a large volume of hot water in the mine workings less than 75 feet from the surface.

Kaolin Deposit

An extensive deposit of kaolin occurs in the vicinity of New York Canyon on the west slope of the Stillwater Range about 25 miles southeast of Lovelock, Pershing County, the nearest railroad point. Very little work has been done on the deposit, and no production has been made.

According to Buwalda,[17] the clay appears to be a sedimentary deposit lying on a series of Paleozoic sandstones and shales, of which it is probably a member. The clay occurs in a stratum 75 to 100 feet thick extending for about 4 miles along the base of the range. In places it is covered to a shallow depth by alluvium, but it could easily be mined in large quantities by open-cut method. The material is white and gritty and contains abundant quartz grains, grains of rutile sparsely disseminated and in places stained with iron oxide.

Fire-tested at 1,150° C., the absorption was 32.4, porosity 48.9, and color white; at 1,300° C. absorption was 28.9, porosity 45.2, and color cream white.

Dixie Marsh

Dixie marsh (also known as Humboldt salt marsh) is about 40 miles north of Frenchman's station on the Lincoln Highway, and it covers an area of nearly 40 square miles in the lowest portion of Dixie Valley between the Stillwater and Clan Alpine Ranges. The valley was known as Osobb Valley by the geologists of the Fortieth Parallel Survey.

The marsh was first exploited for salt in the sixties by John and Wm. Guthrie of Winnemucca, Nev., and associates. Prior to the completion of the Central Pacific R. R. across the State in 1869, considerable quantities of salt were hauled by mule teams to the Humboldt, Reese River, and Comstock districts in Nevada, and some was shipped as far as Silver City, Idaho, for metallurgical use. With the decline of silver milling by the Washoe and Reese River processes, the production of salt ceased and, owing to the isolation of the deposit from consuming centers, no attempts were made to mine salt for other purposes. Some borax was produced from the north end of the marsh in the early seventies.

Several projects were initiated for the recovery of potash from the brines and saline muds in the marsh before and during the World War, but these efforts proved fruitless as far as production of potash is concerned. The Federal Geological Survey tested the salines in Dixie marsh with special reference to potash in 1916, and during the same year prospecting for potash was taken up by the Railroad Valley Co. of Tonopah, Nev. The Railroad Valley Co. sank three holes ranging from 71 to 98 feet in depth, and sampling results of the salts obtained by evaporating the brines showed 28.70 to 38.70 grams of solids per 100 cubic centimeters of material. The percentage of K_2O in the total solids ranged from 0.19 to 0.75 percent, too low to be considered of economic importance. .ne approximate amount of other salts present in the solids is NaCl, 27 percent; Na_2SO_4, 5 percent; and Na_2CO_3, 4 percent.

In December 1917, a group from Fallon, Nev., again took up the Dixie Valley potash project and formed the Nevada Potash Syndicate, but the venture proved unsuccessful. In recent years there has been no activity.

The Dixie marsh was formerly the site of a shallow lake, the evaporation of which produced a mixture of salts including sodium chloride, sodium sulfate, and sodium carbonate, with smaller amounts of sodium borate and potash salts associated with silt and mud. Sodium chloride occurs as an efflorescence in the lowest part of the basin, covering an area of about 9 square miles. The surface salt layer is underlain by a series of salt and saline mud strata to a maximum depth of probably several hundred feet. When the salt deposit was exploited, it was simply hoed into piles and shipped without refining. The borate mineral was chiefly ulexite or "cotton balls" that occurred as aggregates of acicular crystals.

Other Mines and Prospects

In the Table Mountain area on both sides of the Stillwater Range are a number of prospects that were active in the sixties, and a few have made a small production.

About 15 miles north of the I. X. L. district on the west slope of the range is the Marvel district, which was the scene of a little activity about 1911, when a group from San Francisco did some prospecting on several small and badly faulted gold veins near the summit of the range. The veins are in a sedimentary formation that has been intruded by diorite. There is no record of any production.

In the northern part of the Table Mountain district in Pershing County is the Linda-Jo mine, which was discovered are worked by Charles Gilbert in the late 1870's. The economic minerals are lead-silver and a little gold occurring in a quartz vein in limestone. The discoverers are reported to have shipped about $30,000 in ore from shallow surface diggings. Note of this property is contained in an early report by the State minerologist[18] and is as follows:

> Recently some very rich ore has been found in the Linda-Jo mine. Five and one-half tons taken to Winnemucca for treatment gave assays as high as $3,000 per ton, and it is estimated that the whole amount will yield nearly $1,000 per ton. The vein is about 5 feet in width, 1 foot of which yields the rich ore. A shaft has been sunk on the vein to a depth of 16 feet.

Between the Boyer ranch and Dixie mine, a distance of about 12 miles on the east side of the Stillwater Range, are several quartz veins carrying gold. A group of claims owned by Tasker L. Oddie and Fred H. Luetjens was worked in 1938 by lessees, who shipped about 25 tons of ore averaging $40 per ton, Near the quartz veins are irregular feldspathic masses containing aggregates of small crystals of rutile and octahedrite. Other gangue minerals, in addition to the feldspar, are quartz, calcite, and mica. No attempt has been made to prospect the titanium deposits.

On the west side of the range, east of Camp Bolivia, is the almost forgotten Cornish Camp, situated in St. Clair Canyon. In the early seventies, John C. Fall and associates, of Unionville, Nev., did considerable prospecting here on quartz veins carrying silver, but there is no record of any production.

In the Copper Kettle district in Grimes Canyon on the west side of the range, a number of copper claims were located in 1908 by J. R. Bunch, Andrew Robert, and others. In 1917 several carloads of copper ore was shipped, and in 1929 11 tons was shipped from the Copper Kettle mine. Ore occurs near a contact of diorite and altered porphyry and contains chalcocite, cuprite, copper carbonates, and some silver.

In New York Canyon are some old prospects that were active in the 1860's, and it is said that George Hearst had his first Nevada mining experience here. An arrastra was built in the mouth of the canyon by Judge Sam Bonnefield of Winnemucca in the sixties, in which a small amount of ore was treated.

In Fenstonemaker Canyon, also on the west side of the Stillwater Range, are several antimony prospects that were discovered in the early days, but they have been inactive for many years and there is no record of any production.

TOY DISTRICT

The Toy district is in the northwest corner of Churchill County near the Pershing County boundary line. It lies in a group of low hills in the south-western part of the Trinity Range and forms the northwest border of the Humboldt River Valley. The only important property in the district from the viewpoint of past production in the Toy tungsten mine.

According to John T. Reid of Lovelock, the first locations in the district were made in 1885 by Paul Trombley for gold and silver, but no deposits of these metals were found and the claims were soon abandoned. In 1907 Peter Anderson and son, A. M. Anderson, of Lovelock, located a group of four claims for gold. In 1908, lessees, while prospecting the Anderson claims, discovered a heavy white concentrate by panning. Some of the rock was sent to David Atkins, a mining engineer of San Francisco, who at the time was interested in the scheelite deposits at Atolia, Calif. Atkins determined the white concentrate to be scheelite, and the claims were purchased by Atkins and associates, who formed the St. Anthony Mines Co., a subsidiary of the Atolia Mining Co. of Calif. the determination of scheelite in the Toy mine is of especial interest, since it was the first one of the many contact metamorphic deposits of tungsten ore subsequently discovered in the United States. The claims were patented and little work was done until 1915, when the price of tungsten concentrates rose considerably owing to the demand imposed by the World War. In 1915, 20 tons of scheelite ore was mined and shipped to Atolia, Calif., for experimental purposes. In the same year a 75-ton gravity-type concentrator was erected at Fanning siding on the main line of the Southern Pacific R. R, 2-1/2 miles from the mine and 2 miles southwest of Toy siding on the same railroad. The claims were developed and the mill was operated in 1916 and 1917 by the company, and for a short time by lessees in 1918, when the price of tungsten dropped and operations were discontinued. It is said that the first shipment of concentrates from the Fanning mill more than paid for the purchase price of the property, the cost of equipping the mine, and the erection of the concentrator. Since 1918 the property has been inactive. The mill was dismantled in 1921.

Toy Mine

The Toy mine, comprising the Tip Top group of five patented claims owned by the Atolia Mining Co., is 4-1/2 miles by road a little south of west from Toy section house and 20 miles southwest of Lovelock, the county seat of Pershing County. The property is developed by a single-compartment inclined shaft 200 feet deep, several short adits, open-cuts, and other workings totaling about 1,500 feet. The only equipment on the ground is a 50-ton ore bin.

The scheelite occurs in lens-shaped deposits along a limestone-granite contact traceable on the surface along the north side of a small ravine for nearly 3/4 mile. The limestone is thin-bedded, with a prevailing strike of about S. 70° W. and dipping from 50° to 80° SE. Along the contact three ore bodies have been mined, and the width of the stopes, as seen from the surface, ranges from 3 to 10 feet. As a rule, the ore occurs in the limestone, but in places some ore has been mined within the granite for several feet from the contact. The scheelite is fine-grained associated with garnet, quartz, calcite, diopside, and other contact metamorphic minerals stained with iron oxide near the surface from the alteration of pyrite.

The tailings at the old Fanning mill site have been located as a placer claim by Arthur T. Green and associates, of Lovelock. The tailings have been scattered over a considerable area, but probably about 6,000 tons could easily be gathered for retreatment if the scheelite content justified a retreatment operation. Water for former milling operations was obtained from a well at the edge of the Humboldt Sink a short distance from the mill site.

Hardscrabble Claim

The Hardscrabble tungsten claim owned by Herbert Hamlin, of Reno, Nev., is 2 miles west of Toy section house and about the same distance east of the Toy mine. This claim was part of a group located in 1915, when the St. Anthony Mines Co. became active in the district. The claims were allowed to lapse, and the principal claim on which virtually all the work was done was relocated by Hamlin in 1936. At the time of the writer's visit the property was inactive.

Development consists of several shallow shafts, the deepest not more than 20 feet, a short adit, and several open-cuts. From the appearance and extent of the workings, no scheelite ore has been produced. There is no equipment on the property.

The formation is thin-bedded shale striking northwest and dipping 40° NE. The shale is intruded by granite, and along the contact the shale is metamorphosed into a fine-grained aggregate of contact minerals, composed chiefly of garnet and epidote. The scheelite is fine-grained and invisible to the unaided eye.

WHITE CLOUD DISTRICT

The White Cloud district is in the vicinity of the canyon of the same name on the west slope of the Stillwater Range, 33 miles northeast of Stillwater and 35 miles southeast of Lovelock, Pershing County, on the Southern Pacific R. R. The camp of Coppereid is in the central part of the district. According to a news item in an early issue of the Reese River Reveille, a newspaper published at Austin, Nev., the district was discovered by Maj. B. B. Bee and Frederick Smith in 1868, but because of difficulties with the Indians it was not organized until 1869. The first work in the district was done in the early 1870's by John C. Fall of Unionville, Nev., who was one of the prominent early-day mine operators in the State. A small copper smelter which operated for a short time, was erected at the mouth of White Cloud Canyon in the nineties. The Nevada United Mining Co., under the management of J..T. Reid of Lovelock, carried on extensive prospecting operations from 1906 to 1912. Since 1912 there has been no activity. the only production has been several carloads of hand-sorted copper ore from surface workings. One carload shipped by the Nevada United Mining Co. averaged 26 percent copper and contained $1 in gold and 1 ounce of silver to each percent of copper.

Nevada United Mining Co.

The Nevada United Mining Co., controlled by John T. Reid of Lovelock, Nev., owns a group of six patented claims at Coppereid in White Cloud Canyon on the west slope of the Stillwater Range. When the company was active, the number of patented claims totaled 33, but all except six reverted to the county for non- payment of taxes.

The workings comprise an adit 3,050 feet in length, several shorter adits, and scattered surface workings totaling about 5,000 feet. There is no mining equipment on the property.

The prevailing formation is granite porphyry intruded into a sedimentary series composed of limestone, calcareous shales, and several beds of gypsum. The sediments have been metamorphosed by the granite intrusion into hornfels, and small bodies of copper ore of contact metamorphic origin occur in a broad zone of mineralized limestone. The copper is associated with masses of specular hematite with variable strike and dip. Along the contact zone the minerals are garnet, epidote, fluorite, quartz, pyrrhotite, pyrite, sphalerite, and chalcopyrite, in addition to the hematite. Near the surface the primary copper mineral, chalcopyrite, has been altered to malachite and chrysocolla.

According to Reid, the main adit crosscutting metamorphosed limestone and shales has a course of S. 30° E., and at a point 1,300 feet from the portal, a body of disseminated sphalerite about 20 feet in width was exposed. This deposit assayed 4-1/2 percent zinc but carried no gold or silver. The point in the adit where the zinc deposit was found is about 500 feet vertically from the surface. At 2,600 feet from the portal of the adit some lead, zinc, silver ore was penetrated that assayed 5 percent zinc, 4 percent lead, and 3 ounces in silver per ton. No lateral work was done on either of the deposits. At the end of the adit a large mass of iron ore was cut, on the north side of which several feet of copper ore, consisting chiefly of chalcopyrite, was found. The face of the adit is about 1,800 feet vertically beneath the surface. No copper deposits large enough for profitable exploitation were found during the course of prospecting operations. When the adit was driven, the work was hampered considerably by a large flow of water, which during the period of maximum flow amounted to more than 1,000,000 gallons per day.

WONDER DISTRICT

The Wonder district is on the west slope of a southern spur of the Alpine range, sometimes called the Augusta Mountains, in central Churchill County. It is 55 miles east of the town of Fallon, the nearest railroad point; 40 of the 55 miles are over the paved Lincoln Highway and the remaining distance is over fair desert road with an easy grade from the highway to the camp, the rise being approximately 2,000 feet in 15 miles. The elevation of Wonder is 5,500 feet.

The first location in the Wonder District was made in April 1906 by T. J. Stroud on the Jackpot group of claims, and the Nevada Wonder mine was located shortly afterward by Murray Scott, William Mays, and others. The discovery of rich silver-gold ore started a stampede from Fairview that began in May of the same year, and in a few weeks over 1,000 locations were made. The discovery attracted considerable attention, and it was not long before a camp of several thousand people was established. In the first few years of the camp's history, a number of companies were organized, but the bulk of the metal yield was derived from the Nevada Wonder mine, incorporated in Delaware on September 19, 1906. Later this mine was taken over by a group of eastern capitalists, who began a systematic development campaign and in 1913 constructed a 200-ton cyanide mill at the mine. Electric power was brought in from Bishop, Calif., and at the time this transmission line had the distinction of being the longest in the world. The Nevada Wonder Mining Co, controlled, by stock ownership, the claims of the Wonder Extension, reorganized North Star, and Hidden Treasure mining companies, totaling 401 acres, 328 of which were patented. In 1910 water was brought to the camp by a gravity pipe line from Horse Creek, a distance of 10 miles. The company ceased operations in December 1919. After a very profitable history. The total amount of dividends paid was $1,549,002. In 1924 the mine equipment was dismantled and most of it-sold. In 1935 the mine and the equipment remaining was purchased by L. F. Curtis of Reno, Nev. In recent years, mining in the district has been by lessees, largely at the Nevada Wonder mine.

The production of the district from 1907 to 1937, inclusive, was $5,952,764, as shown in table 4.

Nevada Wonder Mine

The Nevada Wonder mine comprises five patented mining claims owned by L. F. Curtis, of Reno, Nev. Development consists of a main three-compartment vertical shaft 1,342 feet deep and an auxiliary shaft 2,000 feet distant sunk to a depth of 800 feet, from which various subshafts and winzes attain a maximum depth of 2,000 feet from the surface. Total underground workings comprise about 8 miles. The lower workings are caved and inaccessible.

Equipment on the property includes a 25-horsepower single-drum gasoline hoist, Rix portable compressor, ore bins, wood head frame, blacksmith shop, and mining tools. In the early part of 1939 several sets of lessees were employed in the upper levels of the mine, and the ore was trucked to the custom milling plant at Westgate for treatment. In the first 6 months of 1938 lessees produced 1,347 tons of ore having a gross smelter value of $43,040.39 or an average of $31.95 per ton.

The country rocks are a complex series of Tertiary eruptives—rhyolite, dacite, andesite, and basalt. The Wonder rhyolite is the principal ore-bearing formation. A number of veins occur,

I. C. 7093

TABLE 4.— Gold, silver, copper, and lead production from Wonder district, Churchill County, Nevada, 1907-37, in terms of recovered metal (Compiled by Charles White Merrill, Mineral Production and Economics Division, Bureau of Mines)

Year	No. of mines	Lode				
		Ore, Short tons	Gold		Silver	
			Fine ounces	Value	Fine ounces	Value
1907.....	3	133	356.38	$7,367	10,993	$7,255
1908.....	6	408	362.13	7,486	79,187	41,969
1909-10..	-	-	-	-	-	-
1911.....	1	9,797	2,476.00	51,183	171,900	91,107
1912.....	1	28,376	7,589.87	156,897	474,316	291,704
1913.....	1	41,870	9,534.00	197,085	699,163	422,294
1914.....	4	50,121	9,715.58	200,839	914,547	505,744
1915.....	2	58,399	9,790.88	202,395	1,175,953	596,208
1916.....	3	58,142	8,955.89	185,135	1,023,288	673,323
1917.....	2	55,804	7,512.74	155,302	816,905	673,130
1918.....	5	49,741	4,883.41	100,949	603,528	603,528
1919.....	5	40,604	5,622.71	116,232	467,283	523,357
1920.....	4	1,218	517.57	10,699	14,505	15,810
1921.....	1	2	1.63	34	2	2
1922.....	2	24	14.89	308	1,755	1,755
1923.....	-	-	-	-	-	-
1924.....	1	1	.38	8	86	58
1925.....	-	-	-	-	-	-
1926.....	1	100	102.67	2,122	902	563
1927-30..	-	-	-	-	-	-
1931.....	3	416	245.20	5,069	13,377	3,879
1932.....	1	200	13.80	285	214	60
1933.....	-	-	-	-	-	-
1934.....	4	1,697	1,173.76	41,023	2,619	1,693
1935.....	2	233	42.76	1,497	14,648	10,528
1936.....	3	364	133.86	4,685	14,009	10,850
1937.....	3	705	294.00	10,290	24,970	19,315
Total.	-	398,355	69,340.11	1,456,890	6,524,150	4,494,132

I. C. 7093

TABLE 4.— Gold, silver, copper, and lead production from Wonder district, Churchill County, Nevada, 1907-37, in terms of recovered metal (cont'd.) (Compiled by Charles White Merrill, Mineral Production and Economics Division, Bureau of Mines)

Year	Lode				Total value	Average recoverable value of ore per ton 1/
	Copper		Lead			
	Pounds	Value	Pounds	Value		
1907......	-	-	-	-	$14,622	$109.94
1908......	-	-	-	-	49,455	121.21
1909-10...	-	-	-	-	-	-
1911......	-	-	-	-	142,290	14.52
1912......	-	-	-	-	448,601	15.81
1913......	-	-	-	-	619,379	14.79
1914......	-	-	62	$2	706,585	14.10
1915......	-	-	-	-	798,603	13.67
1916......	4,564	$1,123	3,350	231	859,812	14.79
1917......	-	-	-	-	828,432	14.85
1918......	1,336	330	602	43	704,850	14.17
1919......	-	-	-	-	639,589	15.75
1920......	-	-	36	3	26,512	21.77
1921......	-	-	-	-	36	18.00
1922......	-	-	-	-	2,063	85.96
1923......	-	-	-	-	-	-
1924......	-	-	-	-	66	66.00
1925......	-	-	-	-	-	-
1926......	-	-	-	-	2,685	26.85
1927-30...	-	-	-	-	-	-
1931......	-	-	-	-	8,948	21.51
1932......	-	-	-	-	345	1.73
1933......	-	-	-	-	-	-
1934......	-	-	270	10	42,726	25.18
1935......	-	-	-	-	12,025	51.61
1936......	-	-	-	-	15,535	42.68
1937......	-	-	-	-	29,605	41.99
Total	5,900	1,453	4,320	289	5,952,764	14.94

1/ Not to be confused with average assay value of ore.

A U.S. Geological Survey geologist captured the view of the Wonder mine and mill, looking north. The electric run 200-ton cyanide plant ran for six years beginning in 1913.

Activity around the Nevada Wonder mine's main shaft consisted of retreiving cars filled with much or waste rock from tunnels and working hundreds of feet below.

most of which contain small deposits of silver-gold ore, but the principal vein from which the major part of the production has been mined is the Nevada Wonder, whose outcrop extends 1-1/4 miles along the strike.

According to Burgess[19] the Nevada Wonder vein lies partly on the contact between rhyolite and the intrusive body of dacite, but toward the north the vein leaves the contact and lies entirely within the rhyolite. The strike is N. 25° W. and the dip is 75° NE. The widths of the oreshoots range from a few feet to a maximum of 30 feet, averaging between 5 and 6 feet. The values are silver and gold in a gangue of quartz, feldspar, and occasional small quantities of fluorite. The gangue is generally stained yellowish-brown with limonite; some of the ore is white. No water is present in the workings, and oxidization extends to the 1,300-foot elevation in the mine. The silver is in the form of argentite and halogen salts, and the gold is both native and combined with argentite. The silver haloids found are embolite, iodobromite, and iodyrite. The ratio of gold to silver by weight, according to production statistics, has been 1 to 94. Oxide of manganese occurs in small dendritic forms, while copper and lead occur only in traces.

Near the surface, where the walls of the vein were firm and stood well, the ore was mined by the shrinkage method. Below the 400 level the walls were less firm, and mining was done by the cut-and-fill system, waste for filling having been obtained from raises driven either into the hanging or foot wall.

FOOTNOTES

1. Lincoln, Francis Church, Mining Districts and Mineral Resources of Nevada, pp. 1-16.

2. Statutes of the Territory of Nevada, 1861, pp. 50-53; see also page 291.

3. Statutes of the State of Nevada, 1903, p. 47.

4. Wickland, Belle, Auditor of the County of Churchill, 22d Annual Report for the Fiscal Year July 1, 1937, to June 30, 1938.

5. Since the foregoing table was prepared, there has been an increase in freight rates and 10 percent of the foregoing rates must be added.

6. Land, Llewellyn L., and Harrington, M. R., Lovelock Cave: University of California Publication in American Archeology and Ethnology, Vol. 25, No. 1, University of California Press, Berkeley, California, 1929, pp. 119-120.

7. Browne, J. Ross, and Taylor, James W., Reports Upon the Mineral Resources of the United States: Gov. Printing Office, 1867, p. 128.

8. Stewart, E. E., Nevada's Mineral Resources: State Printing Office, Carson City, Nev., 1909, p. 110.

9. Mallory, Willard, Antimony Veins at Bernice, Nev.: Min. and Sci. Press, Vol. 112, 1916, p. 556.

10. Greenan, James O., Geology of Fairview, Nev.: Eng. and Min. Jour., Vol. 97, 1914, pp. 791-793.

11. Gale, H. S., Nitrate Deposits: Geol. Survey Bull. 523, 1912, pp. 19-25.

12. Browne, J. Ross, Resources of the States and Territories West of the Rocky Mountains: U. S. Gov. Printing Office, Washington, 1869, pp. 310-311.

13. Russel, I. C., Geological History of Lake Lahontan: Geol. Survey Monograph 11, 1885, pp. 73-80.

14. Knapp, S. A., Occurrence and Treatment of the Carbonate of Soda Deposits in the Great Basin: Min. and Sci. Press, Vol. 77, 1898, p. 448.

15. Newberry, S. B., Mineral Resources of the United States, 1883 and 1884: Geol. Survey, p. 539.

16. Carpenter, Arthur Howe, Boyer Copper Deposits, Nevada: Min. and Sci. Press, Vol. 103, 1911, pp. 804-805.

17. Ries, H., Bayley, W. S., and others, High-Grade Clays of the Eastern United States, With Notes on Some Western Clays: Geol. Survey Bull. 708, pp. 122-123, 1922.

18. Whitehill, H. R., Report of the State Mineralogist of the State of Nevada for the years 1877 and 1888: Carson City, Nev., pp. 65-66.

19. Burgess, J. A., the Halogen Salts at Wonder, Nev.: Econ. Geol., Vol. 12, 1917, pp. 589-593.

MINERAL COUNTY GENERAL INFORMATION

General

Mineral County, in the west-central part of Nevada, was created out of the northern portion of Esmeralda County, and its boundaries were established by an act of the State Legislature approved February 10, 1911.[1] In 1933 certain northern townships and parts of townships were withdrawn from Mineral County and annexed to Lyon County by act of the State Legislature approved March 28, 1933. Before this area was annexed to Lyon County, Mineral County had a land area of 4,019 square miles and a water area of 125 square miles, a total of 4,144 square miles.[2] No accurate figures are available on the area withdrawn from Mineral County in 1933, but it is roughly figured to contain 475 square miles. The water area embraces Walker Lake in the northwestern part of the county.

Figure 1 is a sketch map of the county. The population of the county, according to the census of 1930, was 1,863, about 1,200 of whom reside in Hawthorne, the county seat, and 400 in Mina.

The welfare of the county depends largely on mining[3], as its ranching and agriculture are of little importance. In addition to mining, the only other activity of any consequence is the maintenance of the Naval Ammunition Depot at Hawthorne, which employs a considerable part of the civilian population of Hawthorne in the storage of explosives.

The assessed property valuation in the county for the fiscal year 1935 was $4,521,755.23, and the tax rate for the same was $2.88 per $100, exclusive of special taxes.

Topography

Mineral County has the characteristic physical features of the Great Basin region, comprising a series of approximately parallel mountain ranges with a general northwest-southeast trend and separated by troughlike valleys. The principal ranges are the Wassuk, Gabbs Valley, Gillis, Pilot, and Excelsior Mountains. The strike of the last range is not like the general strike of the others, as it has a general east and west trend. All the mountains are more or less mineralized. The valleys have gently sloping sides and in their lowest portions are filled with a mixture of silt and alkali salts that form the so-called "dry" lakes. The altitude of the lower parts of the valleys ranges from 4,000 to 4,500 feet above sea level.

The mountain ranges are from 6,000 to over 10,000 feet above sea level. The highest mountain in the county is Mount Grant in the Wassuk Range northwest of Hawthorne. Its summit is 11,303 feet above sea level, and it is accessible by automobile road constructed in 1935.

Water Resources

Walker Lake is the only large body of water in the county. It lies between the Wassuk and Gillis Ranges, is 23 miles long, and has a maximum width of 7 miles and maximum depth of 435 feet. Although the lake has no apparent outlet, the water is fresh enough to support trout, bass, carp, and other species of fish. Walker Lake is fed by Walker River and its tributaries, which have their source in the Sierra Nevada Range.

Most of the precipitation is in the form of snow during the winter months. In the valleys the precipitation is approximately 5 inches per year. The amount of precipitation increases progressively with the altitude, and in the higher mountains within this area is appreciably greater. The sides of the mountain ranges are intersected by short, steep canyons, in which small streams are formed by melting snow during the spring run-off. No surface water from this area flows to the sea, and the valleys constitute the drainage reservoirs. Mountain streams supply the water for the town of Hawthorne and the Naval Ammunition Depot, while water for Mina is obtained from springs.

As in all semiarid regions of the southwest, heavy downpours of rain, called cloudbursts, occur occasionally during the summer months. These storms are local in character and bring to the area affected as much rain in a few hours as would normally fall during the whole year.

In some of the mining districts water for milling is scarce. At Sodaville, water for milling is obtained from a number of highly mineralized hot springs. The hot spring water is unsuited for domestic use, but it is used for milling purposes, apparently with satisfactory results. In a number of places within the county are scattered springs and seeps that occur on the slopes of the mountain ranges, but the flows usually are small. In some of the valleys water can probably be obtained from shallow wells in sufficient quantities for milling purposes. The only artesian water in the county is from two wells sunk at the south end of Rhodes Marsh in 1930.

Climate and Vegetation

The climate of Mineral County is characteristic of the high, arid regions in the Great Basin. In the summer the temperature often rises above 90° F. during the day, but the humidity is low, so that this temperature can be endured without discomfort. During the summer months the nights invariably are cool. The winters usually are mild and open, but occasionally freezing temperatures are maintained for short periods. In virtually all the mining districts mentioned in this report mining activity can be carried on the year around without difficulty.

Some kind of vegetation is present over the whole area, with the exception of valley playas, which are filled with a mixture of silt and alkali salts, injurious to vegetation. At the lower altitudes the vegetation consists mainly of sagebrush, and on the mountain slopes stunted growths of juniper, mountain mahogany, and cedars occur. The forest areas contain no sawtimber, and the growth is fit only for firewood.

Power

Most of the mining areas in Mineral County depend upon either Diesel or gasoline engines for power. The only public-service power company operating in the county is the Mineral County Power System, owned and operated by the county. Power is generated at hydroelectric plants in the Sierra Mountains in California, and the transmission line serves the towns of Hawthorne and Mina, and Aurora, Candelaria, Cedar Mountains (Simon and Omco mines), Hawthorne (Lucky Boy Mine), and Silver Star (Silver Dyke mine) mining districts.

The rates for industrial mining and milling power consumption under schedule E, when the consumption exceeds 10,000 kilowatt hours per month, are as follows:

	Cent per kw.-hr.
First 1,000 kw.-hr. per month	0.04
Next 2,000 kw.-hr. per month.	.03
Next 3,000 kw..hr. per month	.02 1/2
Next 4,000 kw.-hr. per month	.02
Next 10,000 kw.-hr. per month	.01 1/2
Over 20,000 kw.-hr. per month	.01 1/4

In addition to the above kw.-hr. charge, there is a readiness to serve or demand charge of $2.50 per month per kw. of 15 minute maximum demand as recorded by meters, provided the monthly minimum payment required shall not be less than $250 per month to cover line and transformer losses so long as any service shall be required of the county.

This demand charge shall be taken as the highest kw. reading for the current month and the 11 months next immediately preceding.

The minimum charge shall be the demand or readiness-to-serve charge, as explained above, of at least $250 per month.

A 10-percent discount is allowed for prompt payment.

Transportation

The Mina-Hazen branch of the Southern Pacific R.R. traverses the central part of the county. This branch line connects with the main line at Hazen. From Mina, the southern terminus of this railroad, the Tonopah & Goldfield R. R., serves the towns of Tonopah and Goldfield. A narrow-gauge railroad, also operated by the Southern Pacific R.R. connects Mina with Keeler, Calif.

An excellent macadam highway running north and south traverses the county. Fair dirt roads branch off the main highway, so that virtually all the districts mentioned in this report are accessible by automobile.

The Southern Pacific R. R. freight rates on ores from the various loading stations in Mineral County to Utah smelters are shown in table 1.

History of Mining

The region comprising the State of Nevada was part of the domain acquired from Mexico by treaty after the close of the Mexican War. Subsequently it was part of the Territory of Utah. By act of Congress approved March 2, 1861, the Territory of Nevada was created, and on October 31, 1864, it was admitted into the Union.

There is no evidence to indicate that the area now included in Mineral County was worked for its mineral deposits while it was a part of Mexico. Gold placers, which are usually the first deposits to be exploited, are relatively scarce in this area, and its mountainous character and vast distance from inhabited centers prevented early settlement.

The first systematic mining began with the discovery of the silver-gold deposits of Aurora on August 26, 1860. This discovery was followed by others, including Candelaria, Garfield, Oneota, Silver Star, and Santa Fe. The period of greatest mining activity was from 1865 to 1875, when these districts attained their maximum production.

In the early seventies great excitement was aroused by the discovery of borax in the salines of Teels, and Rhodes Marsh in Mineral County and Columbus and Fish Lake Marsh in Esmeralda County. Large plants were erected, which were kept constantly at work night and day 8 months of the year. The crude borax was hauled to Wadsworth, Nev., 130 miles distant, by freight teams. In the summer of 1875, 28 teams, 16 horses in each, were engaged in hauling ore and supplies. As the result of the success of the borax operations in Nevada, the principal

I.C.6941   TABLE 1. -- Freight rates on ores from various localities in Mineral County to Salt Lake City smelters.

Value of ore per ton[1]	$10[2]	$15	$20	$30	$40	$50	$60	$70	$80	$90	$100	$150	$200	$250	$300
Schurz	3.20	---	([3] 3.40 ([4] 5.10	4.10 5.80	4.80 6.50	5.50 7.20	6.20 7.90	6.90 8.60	7.00 9.30	8.30 10.00	9.00 10.60	10.70 13.10	10.70 13.10	11.20 13.60	11.20 13.60
Thorne	3.20	---	(3.50 (5.10	4.20 5.80	4.90 6.50	5.60 7.20	6.30 7.90	7.00 8.60	7.70 9.30	8.40 10.00	9.10 10.60	10.70 13.10	10.70 13.10	11.20 13.60	11.20 13.60
Kinkead) Luning)	3.20	---	(3.60 (5.10	4.30 5.80	5.00 6.50	5.70 7.20	6.40 7.90	7.10 8.60	7.80 9.30	8.50 10.00	9.20 10.60	10.70 13.10	10.70 13.10	11.20 13.60	11.20 13.60
Mina	3.20	---	(3.60 (5.10	4.30 5.80	5.00 6.50	5.90 7.20	6.40 7.90	7.10 8.60	7.80 9.30	8.50 10.00	9.20 10.60	10.70 13.10	10.70 13.10	11.20 13.60	11.20 13.60
Sodaville..	---	---	3.90	4.60	5.30	6.00	6.70	7.40	8.10	8.80	9.50	10.70	10.70	11.20	11.20
Rhodes	---	---	5.60	7.00	7.00	8.40	9.10	9.90	10.50	11.10	11.60	13.10	13.10	13.60	13.60
Belleville.	---	---	(3.90 (6.30	4.60 7.00	5.30 7.00	6.00 9.10	6.70 9.60	7.40 10.10	8.10 10.60	8.80 11.10	9.50 11.60	10.70 13.10	10.70 13.10	11.20 13.60	11.20 13.60
Basalt.....	---	(3.90 (6.60	3.90 7.00	4.60 7.00	5.30 7.00	6.00 9.10	6.70 9.60	7.40 10.10	8.10 10.60	8.80 11.10	9.50 11.60	10.70 13.10	10.70 13.10	11.20 13.60	11.20 13.60
Mt. Montgomery	---	4.00	4.00	4.70	5.40	6.10	6.80	7.50	8.20	8.90	9.60	10.70	10.70	11.20	11.20
Queens.....	---	6.90	7.00	7.00	7.00	9.40	9.90	10.40	10.90	11.40	11.90	13.10	13.10	13.60	13.60

[1] The railroad value per ton is the smelter value (less treatment) divided by the total number of tons (wet weight) in the shipment.
[2] Applies on carload minimum weight 100,000 pounds.
[3] First row of figures apply on carload shipments, minimum weight 80,000 pounds.
[4] Second row of figures apply on carload shipments, minimum weight 40,000 pounds.

Mining Districts in Mineral County, Nevada

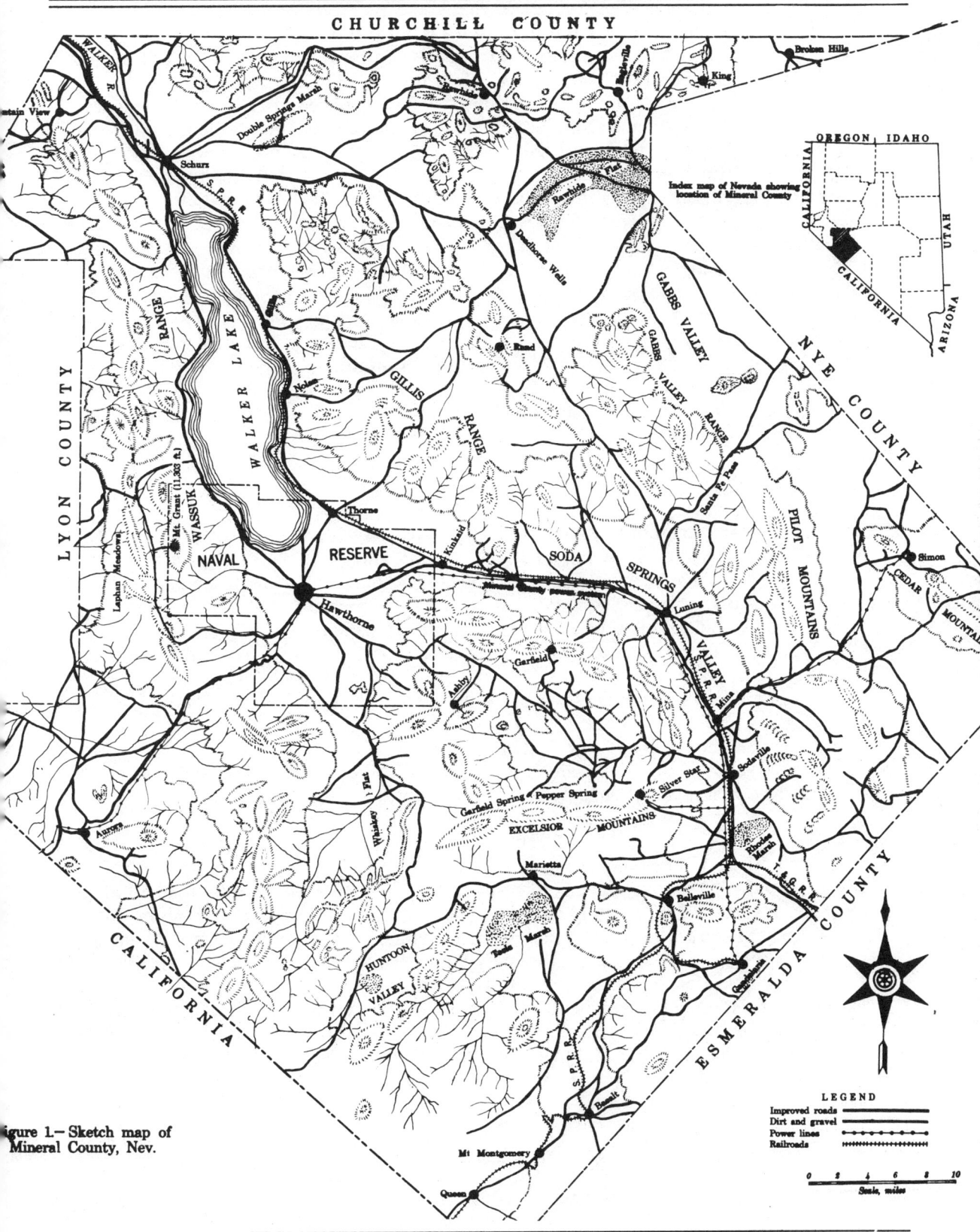

Figure 1.— Sketch map of Mineral County, Nev.

producer, F. M. (Borax) Smith, extended his activities to other borax deposits and eventually obtained control of the world borax market, which he held for about 20 years.

The Carson and Colorado narrow-gage railroad was completed in 1882. With the discovery of bonanza ores of Tonopah and Goldfield, the road was inadequate to handle the traffic, and it was converted to a broad-gage line in 1904.

The mining districts included in this report have been active intermittently since the early days, experiencing alternate periods of prosperity and decline. The last period of intense mining activity in the county occurred during the World War, when the prices of metals rose considerably. Cinnabar deposits were found and worked in the Pilot Mountains, and the tungsten deposits in the Excelsior Mountains began to be exploited. By 1930 the production of metals had reached an all-time low with only $26,699 worth for the entire county. Since 1930 there has been a gradual increase in mining activity by lessees. Since lessees usually are men of small means and hence unable to carry on extensive development, their activities are restricted to the mining of ore of shipping grade that occurs in such a manner that a minimum of dead work is required to recover it. In recent months a number of deposits in the county were being investigated by investors from other States, and the general feeling is that the county will again attain a production comparable to that of former years.

Metal Production

Mineral County is appropriately named, as it contains a greater variety of mineral products in proportion to its size than any other of the 17 counties in the State. The principal minerals produced in the past have been gold, silver, copper, lead, tungsten, and quicksilver. In addition, commercial quantities of placer gold and zinc, in the metal group, and salt, soda, borax, sodium sulphate, clay (bentonite), diatomaceous earth, barite, and andalusite, in the industrial mineral group, have been mined in commercial quantities. Turquoise also has been produced for simiprecious gem stone.

The writer estimates that in October 1936, 200 men were employed in the mining industries in the county.

Mineral production of the county from 1910 to 1934 is shown in table 2. The table does not include tungsten and quicksilver produciton. The production of these two metals for the period given in the table was approximately $1,300,000.

ASHBY DISTRICT

The Ashby district is in an unnamed group of low hills about 20 miles by road a little north of west from Mina. The nearest shipping point is Kinkead Siding, 11 miles to the north.

No water has been developed in the immediate vicinity of the camp; the nearest water supply is probably at Whiskey Springs, approximately 8 miles in an air line to the southwest and at an altitude roughly 1,000 feet lower than that of the camp.

Some prospecting had been done in this area in former years, but no ore was shipped until 1933, when a discovery was made by George A. Ashby of Hawthorne, Nev. The veins are covered with 2 to 20 feet of detrital material and were found by tracing float. Production from 1933 to the time of the writer's visit in June 1936 had been approximately $40,000 worth of shipping ore produced by lessees.

Ashby Gold Mine, Inc.

The principal property in the area is Ashby Gold Mine, Inc. This property comprises six

TABLE 2. — Metal production Mineral County, 1910-34

Year	Lode gold		Placer gold		Silver, lode and placer		Copper	
	Fine ozs.	Value	Fine ozs.	Value	Fine ozs.	Value	Pounds	Value
1910[1]	8,307.78	$171,737	579.77	$11,985	522,987	$282,413	47,375	$6,017
1911	4,696.01	97,075	324.06	6,699	165,808	87,878	12,868	1,609
1912	3,144.04	64,993	81.90	1,693	320,275	196,969	493,719	81,464
1913	6,034.11	124,736	8.22	170	228,832	138,215	769,536	119,278
1914	18,132.69	374,836	259.58	5,366	194,762	107,703	244,536	32,523
1915	36,063.37	745,496	344.82	7,128	292,242	148,167	433,584	75,877
1916	34,264.64	708,313	35.94	743	431,130	283,683	2,986,361	734,645
1917	34,470.82	712,575	20.70	428	407,302	335,617	3,461,969	945,117
1918	27,957.95	577,942	47.55	983	351,760	351,760	2,228,337	550,399
1919	9,204.89	190,282	45.33	937	84,240	94,349	296,815	55,208
1920	4,280.99	88,496	—	—	59,021	64,333	70,886	13,043
1921	5,199.64	107,486	6.58	136	47,561	47,561	6,184	798
1922	981.86	20,297	2.18	45	193,919	193,919	5,063	684
1923	2,928.48	60,537	2.03	42	635,350	520,987	17,942	2,637
1924	1,938.09	40,064	8.37	173	103,956	69,650	1,726	226
1925	2,902.45	59,999	—	—	313,847	217,810	13,282	1,886
1926	1,748.61	36,147	—	—	287,696	179,522	57,054	7,987
1927	2,139.24	44,222	21.33	441	117,490	66,617	48,962	6,414
1928	2,459.00	50,832	—	—	45,890	26,846	22,971	3,308
1929	2,009.22	41,534	—	—	37,167	19,810	56,536	9,950
1930	1,141.70	23,601	—	—	5,307	2,043	6,538	850
1931	1,615.38	33,393	18.29	378	2,429	704	—	—
1932	2,257.47	46,666	42.04	869	7,762	2,189	367	23
1933	1,897.39	39,223	74.61	1,542	11,837	4,143	1,460	93
1934	3,659.33	127,894	39.92	1,395	16,380	10,589	5,028	402
	219,435.15	4,588,376	1,963.22	41,153	4,684,950	3,453,477	11,289,099	2,650,438

[1] Mineral County was part of Esmeralda County prior to 1910.

Continued —

I. C. 6941

TABLE 2. - Metal production Mineral County, 1910-34 (Continued)

Year	Lead		Zinc		Total value
	Pounds	Value	Pounds	Value	
1910[1]/	711,517	$31,307	--	--	$503,459
1911	184,662	8,310	--	--	201,571
1912	298,598	13,437	--	--	358,556
1913	59,632	2,624	--	--	385,023
1914	60,800	2,371	--	--	522,799
1915	221,047	10,389	--	--	987,057
1916	270,734	18,680	--	--	1,746,064
1917	357,876	30,777	--	--	2,024,514
1918	171,660	12,188	--	--	1,493,272
1919	65,204	3,456	--	--	344,232
1920	158,555	12,684	--	--	178,556
1921	111,500	5,017	--	--	160,998
1922	2,056,098	113,085	1,224,000	$69,768	397,798
1923	3,124,176	218,692	2,320,053	157,763	960,658
1924	96,068	7,685	--	--	117,798
1925	740,007	64,381	255,000	19,380	363,456
1926	1,693,400	135,472	910,125	68,259	427,387
1927	654,930	41,261	285,406	18,266	177,221
1928	51,000	2,958	--	--	83,944
1929	40,036	2,522	1,113	73	73,889
1930	4,100	205	--	--	26,699
1931	--	--	--	--	34,475
1932	17,478	524	--	--	50,271
1933	26,603	984	--	--	45,985
1934	9,497	351	--	--	140,631
	11,185,178	739,360	4,995,697	333,509	11,806,313

[1]/ Mineral County was part of Esmeralda County prior to 1910.

ASHBY DISTRICT

unpatented claims owned by J. H. Miller, H. S. Babbitt, and George A. Ashby, all of Hawthorne, Nev. Development work consists of 10 vertical shafts, the deepest of which is 200 feet. Underground workings total about 2,000 feet. These shafts have been sunk by lessees, and several are equipped with small gasoline hoists and air compressors. In June 1936, three sets of lessees were working on the property and combined production averaged about 1 car of ore per month. The royalty paid by lessees varies with the grade of ore shipped and is as follows:

Value of ore	Royalty, percent
$50 or less	10
$50 to $75	15
$75 to $100	20
$100 or more	25

The royalty payments are based on net smelter returns after hauling, freight, and smelter charges have been deducted. Leases are given for 18-month periods and on blocks 150 feet long on the strike of the vein and 200 feet in width.

A series of veins varying in width from 6 inches to 6 feet occurs in an altered andesite formation cut by aplite dikes. The principal values are in gold with minor amounts of silver.

A shipment of ore made from the Armanko-Cafferata lease to the U. S. Smelting and Refining Co. April 1, 1935, provided the following information:

Metal Quotation:
    Au   34.9125
    Ag   .77
Settlement assay:
    Au   6.53 oz.
    Ag   5.0 do
    Cu   .3 percent
    Pb   .6 do
    Insol.   66.0 do
    Fe   12.4 do
    S   .9 do
    CaO   .5 do

Metal payment:
Au, 100 percent at 32.40		$211.58	
Ag, 100 percent at .77		3.85	
		215.43	
Less 0.5 oz. Ag		.39	
		215.04	
Treatment charge		6.20	
Net value per ton		208.84	
Wet weight	73,400		
Sacks (666)	392		
	73,008		
Less Moisture 6.2%	4,526		
	68,482	equals 34.241 tons at $208.84	$7,150.89
Freight, at $10.70 per ton		$392.69	
Sacks returned		2.68	
Hauling, at $1.75 per ton		64.22	
Royalty, 15 percent		1,003.69	1,463.28
		Net proceeds	5,687.61

The royalty of 15 percent on this shipment of ore is less than the royalty payment shown in the table, because in this Particular case the lessees agreed to do a certain amount of "dead" work in return for a reduction in the royalty.

The haul to Kinkead Siding 11 miles from the property is done under contract at $1.75 per ton. The above shipment represents the best shipment of ore from the district. The average value of the ore has been $25 per ton.

According to J. H. Miller, the material in the dumps will average $5 per ton. If the ore is screened through a 1/2-inch screen the undersize will average $12 per ton. The fines constitute about 25 percent of the dump material. The mine dumps contain, roughly, several thousand tons.

With the present stage of development in the district, there is not sufficient tonnage of ore in sight to justify the expense of erecting a mill.

AURORA DISTRICT

Aurora, known in the early days as the Esmeralda district, is in western Mineral County, 3 miles east of the California-Nevada boundary line and at the head of Aurora Canyon, which is tributary to Bodie Canyon. Aurora is 30 miles by automobile road southwest of Hawthorne via Fletcher's Station. The altitude of Aurora is approximately 7,400 feet above sea level. The area is quite mountainous, and during the winter months snowfalls are sometimes heavy so that the camp is inaccessible by automobile. Mining, however, can be carried on the year round.

Aurora is supplied with electric energy from the high-tension transmission line of the Mineral County Power System.

Water for milling is available at the Prospectus drainage tunnel at Gregory Flat and at Tamarack Springs, which are reported to have a flow of 4 miner's inches. Tamarack Springs are at an elevation of 9,000 feet and, according to a survey, the water can be carried to the town by a gravity pipe line 8,300 feet long.

Veins carrying gold and silver were discovered here on August 26, 1860, by E. R. Hicks and party while they were hunting for game. Shortly after the discovery a spectacular rush ensued, and the camp of Esmeralda was established on Gregory Flats. Later, the town of Aurora was established several mines distant from Esmeralda camp. By an act of the First Territorial Legislature, November 25, 1861, Esmeralda County, named after the mining district, was made one of the nine original counties of Nevada, with Aurora as the county seat. The town of Aurora was substantially built, and a number of the houses and stores were constructed of brick and masonry.[4] At present most of the buildings are in ruins, and the general atmosphere of the camp is one of neglect and decay.

In 1864, Aurora had a population of nearly 10,000, but by 1869 the bonanza ore near the surface became exhausted and a considerable part of the population moved to Virginia City. The mines, however, continued to produce up to 1882.

In the early days as many as 17 mills were operating in the district at one time. These mills employed stamps for crushing and the Washoe pan process for recovering the values. Compared with modern processes, the recovery in these early-day mills was low, and the tailings that were available were subsequently cyanided in 1901 and 1902.

In 1912 the Aurora Consolidated Mines Co. was incorporated. This property was purchased in 1914 by the Goldfield Consolidated Mines Co., which erected a 500-ton mill equipped with 40 stamps, each weighing 1,750 pounds. Primary crushing was done with the stamps and fine grinding with 3-tube mills. Countercurrent cyanidation was employed to recover the values. After about 3 years operation the mill was dismantled and the equipment sold.

At present there are two small mills at Aurora. One is a 10-stamp affair erected in 1912 and owned by W. J. McKeough of Aurora. In 1935, the mill was operated for a short time by the Western Consolidated Mines Co. This company ran into financial difficulties and closed down. Mill equipment consists of two 5-stamp batteries (1,050-pound stamps), a ball mill 5 feet by 4

A rich mining camp of the 1860's and 1870's, Aurora revived early in this century when stores, newspapers and the post office reopened. The famed Esmeralda Hotel dominates the left flank of town, while on the right are saloons, stores and lodging houses. Below is a general view of Aurora looking north into Aurora Canyon.

1/2 feet, 2 amalgamation plates, a Dorr simplex classifier and 2 Groch flotation cells. The other mill is a Kincaid mill with a capacity of 2 tons per day and is owned by Freid Walker of Aurora.

In recent years mining activity has been confined to small leasing operations, and the small amounts of ore produced were either shipped to smelters or milled locally. In June 1936, a crew of four men, under the direction of Walter Trent, was employed in sampling old workings.

As in all early-day mining camps, the records of the production of Aurora are incomplete. According to the records of the Wells Fargo Co., the bullion shipped through them up to 1869 had a value of $27,000,000. In addition, there is a record of $2,365,969 shipped without insurance; therefore, it can be stated safely that the output of the district between the years 1861 and 1869 was about $30,000,000. This sum appears conservative if statements of production from some of the high-grade stopes are considered. From 1910 to 1920 the district produced $1,882,861 in gold and 128,808 ounces of silver, valued in all at $1,974,290, according to Mineral Resources of the United States. Production from 1915 to 1917 made by the Goldfield Consolidated Mines Co. is shown in the following table:

TABLE 3. - Mineral production of Aurora District, Mineral County, Nev., 1915-17, inclusive.

Year	Ore and tailings, tons	Average value, per ton	Total value
1915	141,421	$3.734	$528,097.00
1916	173,270	2.892	501,041.14
1917	175,477	2.315	406,277.52
Total	490,168	2.93	1,435,415.66

The mineralized belt is about 2 miles long and 1 mile wide. Total workings comprise about 20 miles. The deepest shaft is the Del Monte, which is 900 feet deep. The depths of other important shafts are: The Gladiator, 450 feet; the Monarch, 400 feet; the Durand, 500 feet; and the Junietta, 450 feet. The Prospectus tunnel near the former mill site of the Goldfield Consolidated Mines is 1 1/4 miles long.

The principal holdings in the area comprise a large group of patented claims owned by the Goldfield Consolidated Mines Co. Other claims in the district are owned by the West End Mines Co., the Aurora Consolidated Mines Co., and W. J. McKeough of Aurora. In addition, individuals or groups of individuals own from 2 to 7 claims each. Many of the claims are small, being 1,500 feet long and 200 feet or less wide.

The formations in the Aurora area are volcanic, probably of Tertiary age, which have flowed out on a basement of porphyritic granite. The oldest flows are latites, with associated andesites, on top of which is a rhyolite flow; above the rhyolite is a flow of basalt.

The ore deposits are quartz veins that fill fissures in the latite and andesite. Some of these are simple quartz veins that range from 2 to 30 feet in width, while others are made up of an interlacing network of small veins. The vein filling is principally fine-grained white quartz occurring in hands and small druses. The veins vary in strike, but in general the strike is northeast-southwest with dips of 45° to 50° to the south.

The rich ores are characterized by irregular streaks made up of quartz, adularia, tetrahedrite, pyrite, chalcopyrite, and a soft bluish green material supposed to be a combination of gold, and possibly silver, with selenium. Free gold is present in the richest ores. The values are principally gold, with some silver.

In conclusion, it can be stated that the general impression gained by the writer in a brief visit to the camp is that the district may be worth investigating from the viewpoint of working the properties on a leasing basis. The area contains numerous narrow interlacing veins that are ore-bearing over a large area. A condition such as this in an old camp from which the cream has been skimmed naturally lends itself to the leasing system, whereby a number of men, working separately or in groups of two or three, can mine ore from comparatively small veins unsuited for mass production, provided metal prices are favorable, an equitable royalty schedule and fair milling rates are established, and the lessees can have their ore treated in a local custom plant, which they themselves cannot afford to erect. Some of the best ground could be reserved for company operations. The writer was informed that a custom mill equipped with 20 stamps operated in the district from 1906 to 1911. This mill was erected by J. S. Cain and employed amalgamation with subsequent cyanide leaching of the amalgamation tailings. Ore was ground to 30-mesh. The mill ran entirely on custom ore and had a capacity of 60 tons per day. The increased price for gold is an important factor in considering an investigation of the possibilities of the district.

The writer has been informed that a number of the old mine dumps carry values from $6 to 11 per ton and that approximately 35,000 tons of such material is available in 12 different dumps. This information would have to be checked by thorough sampling.

BASALT DISTRICT

Basalt is a station on the narrow-gage railroad that operates between Mina, Nevada, and Keeler, Calif. It is 22 miles west of Coaldale, Esmeralda County, Nev., and is accessible from this place by an excellent highway.

A deposit of diatomaceous earth approximately 3 miles long and 1/2 mile wide occurs in the vicinity of Basalt. This deposit has been prospected by a number of shafts and open cuts. The deepest shaft is 135 feet and the bottom is still in diatomaceous earth. The depth of the overburden, which consists of desert wash and a few basalt boulders, is not more than a few feet. Although this occurrence of diatomaceous earth has been known since 1905, only small shipments have been made. The last shipments were made in 1927 and 1928, when about 5,000 tons of the material was shipped to Los Angeles for filtering and construction purposes. For all practical purposes the supply of diatomaceous earth is un- limited.

Somerville Group

Robert D. Somerville of Basalt owns a group of 16 claims of 40 acres each, covering part of the deposit. The earth could be mined easily with a power shovel, and, in one place the distance to the railroad is not more than 1/4 mile. When shipment is made by rail, the material would have to be transferred from narrow-gage to standard-gage cars either at Mina, Nev., or Keeler, Calif.

On the Somerville ground, about 2 miles east of Basalt, is a deposit of pumicite. This deposit was shown to the writer and reported to be alunite. microscopic examination of the material, however, shows that it is composed entirely of fine grains of silica. The deposit is 10 feet wide and dips 60 degrees; both walls are diatomaceous earth. It has been prospected by an open cut about 15 feet long and a maximum of 10 feet deep.

Diatom Company

A company called the Diatom Company, controlled by Langlois Brothers, 717 South San Pedro Street, Los Angeles, Calif., also owns 16 claims covering the diatomaceous earth. Most of the production of the diatomaceous earth has been made by this company.

BELL DISTRICT

The Bell, also known as the Cedar Mountain District, is in the Cedar Mountain Range in eastern Mineral County near the Nye County border. It includes the camps of Omco, Simon, and Copper Contact. Simon is 22 miles by road northeast of Mina, and Omco lies 4 miles north of Simon. The principal properties in this area are the Simon and Omco mines.

Simon Silver-Lead Mines, Inc.

The Simon mine was discovered in 1879, at which time small quantities of lead ore mined from the gossan were shipped. Its importance was not discovered, however, until 1919, when silver-bearing lead-zinc ores were discovered in the sulphide zone below the gossan that had been prospected in 1879.

In 1921 the Simon Silver-Lead Mines Co. erected a 100-ton flotation mill at the mine. In 1923 this company was reorganized under its present name and the mill enlarged to handle 250 tons per day. Up to 1927 the mill, operating at four different periods, had treated 93,000 tons of ore and produced 6,258 tons of lead and 5,311 tons of zinc concentrates having a gross smelter value of $741,278. The mill closed in January 1927.

Property consists of a contiguous group of patented and unpatented claims amounting to 600 acres. In the acquisition of this acreage the present company absorbed 7 smaller companies.

The Simon mine is opened by a 3-compartment vertical shaft 800 feet deep and a winze sunk to a depth of 200 feet from the 800-foot level. Total underground workings comprise in the neighborhood of 25,000 feet. The shaft and underground workings are reported to be in good shape, but, recently the mine has been allowed to fill with water to the 450-foot level.

Mining equipment includes a 150-horsepower hoist, a compressor, a drill sharpener, pumps, lighting plant, and camp buildings.

Mill equipment includes a No. 5 Allis Chalmers gyratory crusher, a No. 3 Kennedy gyratory crusher, a 6- by 10-foot Allis Chalmers tube mill, a 62 1/2-inch Marcy ball mill, 2 Dorr drag classifiers, 26 18-inch sub-A mineral separation flotation cells, 2 5- by 6-foot Oliver filters, and 4 Dorr thickeners.

Purchased power is available at the property.

The geology of the district has been described by Knopf[5], who stated that the ore bodies are replacement deposits in limestone localized along an alaskite dike about 30 feet wide. The minerals in the sulphide zone are argentiferous galena and zinc blende associated with subordinate amounts of pyrite and chalcopyrite in a gangue of calcite and limestone.

Considerable development work was done after the mill closed in 1927, and, according to J. H. Simpson, manager of the property, about 100,000 tons of ore have been blocked out. The average value of this ore, as determined from development samples, was stated to be as follows:

Au	0.04 oz.
Ag	9.0 oz.
Pb	9.0 percent
Zn	8.5 percent
Cu	.2 to 3 percent

The intense faulting and other complicated geological features of the mine, together with the grade of the ore, has made it impossible to operate profitably under the conditions that prevailed in the base-metal market in recent years.

Omco Mine

The Omco mine is 25 miles northeast of Mina at the north end of the Cedar Range.

The Royal George group of nine claims was located in 1915 by James P. Nelson. The property was sold shortly after to San Francisco interests that organized the Olympic Mines Co. Omco is an abbreviation of the name of this company.

The Olympic Mines Co. erected a 70-ton cyanide mill in 1917. This mill burned down in 1919 and another of 80 tons capacity was built in 1920, which closed in 1921. The mine has been operated at various times by lessees, and in 1929 it was sold at a tax sale. On December 20, 1932, the mine workings were caved by a severe earthquake that occurred in this vicinity. In 1936, the mine and mill were being rehabilitated by J. H. Simpson and associates of Simon.

Production, principally by the Olympic Mines Co., has been about $700,000, principally in gold. About 35,000 tons of ore, at from $15 to $20 per ton, were treated in the two mills, and in addition considerable ore was shipped.

The Omco mine is developed by a shaft 225 feet deep, inclined 43 degrees. Total lateral workings comprise about 3,000 feet.

The Omco mill equipment includes a Hendy crusher, a 6- by 1 1/2-foot Hendy tube mill, a 5- by 6-foot Hendy ball mill, a Dorr duplex drag classifier, two 12- by 14-foot Oliver filters, two 12- by 18-foot Dorr agitation tanks, three 12- by 24-foot Dorr thickener tanks, pumps, refinery and other cyaniding equipment.

An extraction of 93 percent is reported to have been made in milling. The mill-tailings dump, which contains about 35,000 tons, has been sampled several times, but the sampling results are not available.

The mine and mill are served with power by a branch transmission line from Simon.

In 1936 a 2-inch-diameter pipe line 4 miles long was laid to carry water from the collar of the Simon shaft to Omco by gravity.

The Olympic vein occurs in rhyolite and trachyte covered in places with tuff. The vein averages 4 feet in width and consists of quartz and more or less silicified rhyolite. Values are chiefly in gold, which is invisible, and the best ore is chalky in appearance with no sign of mineralization. The vein has a dip of 43 degrees to the 100-foot level, from which point it becomes flatter to the 150-foot level. To the west the vein has been cut by a fault, and the displaced segment has been lost or eroded.

Golden Mile Group

The Golden Mile group of four unpatented claims owned by J. H. Walsh of Mina is at Copper Contact, 22 miles by road northeast of Mina and several miles south of Simon.

This property was discovered in 1902 by Jesse Workman. In 1935 it was worked under bond and lease to Henry Ott of Reno and associates, who shipped approximately 4,000 tons of ore and relinquished the option. Total production of shipping ore is reported to be about 10,000 tons.

Development consists of three tunnels and other workings, totaling about 1,000 feet. The ore mined has been taken from an open-cut several hundred feet long, 25 feet deep, and 30 feet wide.

In 1936, J. H. Walsh mined several carloads of ore by hand methods. The smelter returns on a carload of ore shipped by Walsh to the American Smelting and Refining Co. smelter at Salt Lake City on May 12, 1936, furnished the following data:

Metal quotation:
    Au   34.9125
    Ag   .77
    Cu   9.15 less 2.525 per pound.

Settlement assay:
- Au — .345 oz.
- Ag — 1.15 oz.
- Cu — 1.66 percent
- Insol — 14.3 percent
- S — .5 percent
- As — .3 percent
- Fe — 16.4 percent
- CaO — 19.3 percent

Metal payment:
Au at $31.81825	$10.98
Ag 95% at .77	.84
Cu less 0.4% at 1.325	1.67
	13.49

Treatment charge:
Base rate	1.35	
Penalties and premiums	1.56	
	2.91	2.91
Net value per ton		10.58

Wet weight	104,280 pounds		
Less 4% moisture	4,172 pounds		
Dry weight 100,108 pounds equal 50.		054 tons at $10.58	$529.57
Freight	$3.60 per ton	187.70	
Hauling	$2.50 per ton	130.35	
Assaying		4.00	
		322.05	322.05
Net Proceeds			207.52

The deposit is a flat-lying vein in magnesian limestone near a granitic intrusive. Values are chiefly in gold with minor amounts of silver and copper. The gangue consists mainly of iron oxide and calcite. An analysis of the ore made by the Union Assay office on 2 carloads shipped to the smelter was as follows:

Au	0.26 oz.
Ag	1.2 oz.
Pb	None
Cu	2.06 percent
Insol	14.2 percent
Zn	.6 percent
S	.2 percent
Fe	33.1 percent
Lime	9.0 percent
Sb	None
As	Trace
$MnCO_3$	Balance

From the analysis, this is an unusual and complex type of gold ore. Metallurgical tests have not been made to determine whether or not this type of ore can be treated by flotation.

Clay Peters Group

The Clay Peters group of three unpatented claims owned by Mrs. Clay Peters of Los Angeles, Calif., is 12 miles northeast of Mina.

Property was last worked in 1935 by William Myers of Omco and two partners, who shipped 14 cars of ore that averaged $18 per ton. Royalty paid by Myers was 15 percent of the net smelter returns. The trucking cost to Mina was $2.25 per ton. Property was idle in 1936.

Development consists of a shaft 240 feet deep and some lateral workings. Equipment includes an Ingersoll Rand Imperial Type 12 compressor driven by a Waukesha gasoline engine. The hoist has been removed from the property.

The orebody is an irregular deposit in limestone. Values are chiefly in gold, with small amounts of lead and copper in a siliceous gangue.

Harvey-Taylor Group

The Harvey-Taylor group of five patented claims owned by J. A. Ashby of Hawthorne, Nev., is 20 miles by road northeast of Mina and 1 1/2 miles west of Simon. A small amount of shipping ore has been produced in the past by lessees. In 1936 the property was idle.

Development work consists of two tunnels, 100 and 165 feet in length, driven on the vein, and two winzes, one 100 feet and the other 150 feet deep. There is no equipment on the property.

Water for milling is available from Storey Springs in the vicinity. The estimated flow from these springs is 12 gallons per minute.

Formation is andesite. The vein has a width of 2 feet on the surface and 4 to 6 feet at the tunnel levels. Vein is said to be exposed for a length of 650 feet, and the dip averages 50°.

Values are in gold and silver. The gold is largely free-milling. According to Ashby, property has been sampled by several engineers and approximately 30,000 tons of ore averaging $5 per ton, at current metal prices, are reasonably assured.

Finger Rock Quicksilver Mining Company, Inc.

The Finger Rock Quicksilver Mining Co., Inc., Kenneth D. Holland, president, is a small stock company recently formed by Los Angeles people. The company owns a group of three unpatented claims several miles northerly from Omco. The property is 26 miles northeast of Luning, Nev., by way of Santa Fe Pass.

This property has produced 5 flasks of quicksilver. In 1936, four men were employed in driving a tunnel to explore the deposit at depth.

Development work consists of a tunnel 250 feet long and a shaft 50 feet deep. Equipment includes one-drill compressor and a 25-ton capacity Cottrell furnace recently installed.

A shear zone contains cinnabar in rhyolite.

Diatomaceous Earth

About 4 miles easterly from Copper Contact, near the Mineral-Nye County boundary line, diatomaceous earth is to be found. The nearest shipping point is Mina, about 25 miles from the deposit.

About 1925, a company, called The Nature Products Co., held over 200 acres of diatomaceous-earth lands in this area and is reported to have shipped 3 carloads to Reno for the production of tooth powder and dental cream.

The deposit is exposed on the surface over a considerable area and the material is said to be of good quality.

Although this deposit could supply large quantities of diatomaceous earth for the building trades, the low price paid for such material and distant markets makes it difficult to compete with other sources of supply.

BROKEN HILLS DISTRICT

The Broken Hills district is in the northeastern corner of Mineral County in a group of low hills between the Fairview and Ellsworth Ranges. It is accessible from Fallon which is on the main line of the Southern Pacific R. R. and 70 miles to the northwest.

Silver-lead ore was discovered here in 1913 by James Stratford and Joseph Arthur. Stratford and Arthur worked the property until 1920, when it was sold for $75,000 to a company called the Broken Hills Silver Corporation, financed by George Graham Rice's Fidelity Finance and Funding Co. This company ran into difficulties and a reorganization was effected in 1921. After the reorganization, some ore from the Broken Hills mine was treated in a 50-ton cyanide mill at Bruner 12 miles distant. Several other small companies were organized in the twenties to work properties in this district, but all these companies have passed out of existence.

The original owners are reported to have shipped ore to the value of $75,000. Lessees have made intermittent shipments of ore totaling about $15,000. Total production of ore probably exceeds $100,000.

The ore occurs in narrow, steeply dipping veins in andesite. Two veins have been found on the property, formerly owned by the Broken Hills Silver Corporation. The vein filling consists of crushed and altered andesite with minor amounts of quartz. The ore contains lead, silver, some zinc, and a little gold. Virtually all the ore mined has been taken from the oxidized zone, which extends to a depth of 150 feet from the surface.

Broken Hills Mine

Ground formerly owned by the Broken Hills and Belmont companies was relocated in 1936 by George M. Lerchen, of Broken Hills, and associates. The previous owners neither filed claim for exemption of assessment work nor did any assessment work, so the ground was relocated by Lerchen and associates. The surface equipment was bought in at a tax sale. The Lerchen property comprises four unpatented claims.

Development work consists of Broken Hills shaft, 600 feet deep; Belmont shaft, several hundred feet deep; and approximately 6,000 feet of underground workings. All mining equipment except a 15-horsepower Fairbanks Morse hoist at the Broken Hills shaft has been removed.

In October 1936, Lerchen and partner were working on a vein on the 150-foot level from the Broken Hills shaft. The width of the vein ranged from 18 inches to 2 feet, and the owners had mined 15 tons of ore reported to average 100 ounces of silver. Mining was done by hand. According to Lerchen's sampling, the Broken Hills ore dump, which contained an estimated 10,000 tons, will average 15 ounces of silver. In addition, approximately 10,000 tons of the same grade of ore are said to be available in the Broken Hills workings.

The nearest water supply is at Halley's well in Lodi Valley, 10 miles distant. The well is reported to be 140 feet deep.

The cost of trucking ore to Fallon for shipment is $4 per ton on contract; 25 miles of this haul is over desert road and 45 miles over the Lincoln Highway.

Silver Trailer Group

The Silver Trailer group of five unpatented claims owned by V. S. Baxter of Fallon adjoins the Broken Hills Mine. This property is developed by a 100-foot shaft and about 1,000 feet of lateral workings.

According to Baxter, the only production from this property is 1 carload of ore shipped several years ago. This ore averaged 80 ounces of silver and 22 percent lead per ton. Property has been idle for several years.

Baxter Mine

The Fluorspar group of eight unpatented claims owned by Vet. S. Baxter, of Fallon, is 6 miles west of Broken Hills. The nearest shipping point is Fallon, on the main line of the Southern Pacific R. R., 70 miles northwest. This fluorspar deposit was discovered and located by Baxter in 1922; since its discovery, intermittent shipments totaling approximately 3,500 tons have been made. Fluorspar is used for fluxing in the steel industry on the Pacific coast.

No attempt has been made to concentrate the fluorspar other than by selective mining and hand sorting. No water is available in the immediate vicinity of the mine, but probably water could be developed within a radius of 10 miles.

Development work comprises three shafts (the deepest of which is 110 feet) and underground workings totaling about 1,000 feet.

The fluorspar is in veins in a formation said to be andesite. The main vein is traceable on the surface for a distance of at least 3,000 feet. The strike of the vein system is about east and west, and the dip is 55° to the north. The width ranges from a few inches to a maximum of 10 feet. The foot-wall of the deposit is a well-defined fault, but in the hanging wall a number of stringers of fluorspar, ranging from a few inches to 1 foot in width, branch off from the main vein. The chances for finding additional parallel veins either in the footwall or hanging wall of the deposit are favorable.

Mining has been confined to the production of shipping ore whenever there has been a market for the product. At the time of the writer's visit in October 1936, two men were employed in working the deposit on contract. Drilling is done by hand and the open-stope method of mining is employed. A 12-horsepower Fairbanks-Morse gasoline engine and 800-pound bucket are used for hoisting.

A high-grade product is produced by selective mining and hand sorting. Pieces of waste shot down with the ore are sorted by hand. The minimum width mined averages about 18 inches. The shipping product contains an average of about 95 percent fluorspar, 1 percent silica, and smaller amounts of iron, manganese, and aluminum.

CANDELARIA DISTRICT

The Candelaria district also known as the Columbus district, is in the Candelaria Mountains at an altitude of 5,665 feet above sea level, 22 miles south of Mina by automobile road. Silver veins were discovered here in 1863 by a party of Spaniards. The Northern Belle, the oldest and most productive mine, was located in 1864 and in the same year was acquired by Messrs. Bateman, Allen, and Holmes. Subsequently this property was abandoned.to be relocated in 1870 by A. J. Holmes. In 1873, a company was incorporated to work the Northern Belle property, and in the same year a 20-stamp mill and 3 White furnaces were erected at Belleville 8 miles from the mine. With the successful development of the Northern Belle property, attention was attracted to the district, and it became the most productive silver camp in Esmeralda County and one of the foremost in Nevada. The 20-stamp mill erected at Belleville was equipped with 10 pans and 6 settlers for treating the ore by the Washoe process (amalgamation in pans heated by steam).

In 1876, a second 20-stamp mill, equipped with 12 pans and 6 settlers, was erected by the Northern Belle Co. at Belleville. The two mills had a combined capacity of 120 tons of ore per day. Power was generated with steam, pine wood being used for fuel. An old report states that the two mills required 1,000 cords of pine wood per month. Other companies erected mills at

Columbus and Sodaville. The value of the ore treated in the early days averaged from $45 to $60 per ton.

In 1883, the Holmes Mining Co., whose property adjoined the Northern Belle, sued the latter company for trespass and sought compensation for ore extracted from its ground. The court awarded the Holmes Mining Co. $360,000 damages, and the Northern Belle Mining Co. mine and mills were sold by the United States marshal in 1884 to the Holmes Mining Co.

As the bonanza ore of the early days became exhausted, the camp declined and fell into decay. The last important revival of activity in the district was in 1919 when the Candelaria Mines Co. was organized. This company was a consolidation of the most important mines, including the Argentum, Mount Diablo, and other properties of less importance.

In 1882, water was brought to Candelaria by gravity from the White Mountains. The pipe line has a diameter of 4 to 5 inches and is 27 miles long. It is still in serviceable condition.

From 1913 to 1918 about 125,000 tons of old tailings at Belleville were re-treated in a 120-ton cyanide plant.

The total production of the district in the early days is reported to have been $20,000,000. According to a report of the Mint,[6] by the end of 1883 the total amount of silver bullion aggregated $10,000,000, out of which about $2,000,000 had been paid as dividends.

From 1903 to the present time production has been in excess of $1,000,000.

Argentum Mining Co.

The Argentum Mining Co. property includes the Northern Belle and Holmes mines and other mineral acreage in the Candelaria district. Fred G. Gruby, 241 Sacramento Street, San Francisco, Calif., is the western representative of this company. In 1918, the Argentum holdings were leased to the Candelaria Mines Co. Lease was canceled in 1927 for failure to comply with its provisions. In 1922, the Candelaria Mines Co. erected a 300-ton cyanide plant, which operated at intervals until September 1925. A large proportion of the ore treated consisted of tailings and material from mine dumps. Considerable difficulty was encountered in the metallurgy. This mill has been dismantled.

Property is developed by the Northern Belle and Argentum shafts, the latter 1,365 feet deep, and many miles of underground workings.

According to Knopf[7] the rocks of the district consist of a steeply dipping series of cherts, argillites, and felsites intruded by peridotite or allied rock and quartz monzonite porphyry. Resting uncomfortably on this group of older rocks is a series of Tertiary volcanics, mainly rhyolite, lavas, and tuffs capped in places by basalt flows.

The ore deposits are highly oxidized manganiferous silver veins mostly several hundred feet in length and a few feet wide, broken up by complex fissures. No silver minerals are visible, and the value of the ore can be determined only by assay. The amount of gold in the ore is of minor im- portance. The veins are fairly persistent and dip at high angles. The deepest workings are 1,365 feet vertically below surface, and at this depth water level has not been reached. There is said to be little hope at depth, either in grade or quantity of ore.

According to Fred G. Gruby, several reports on the property made by prominent engineers give estimates of ore reserves in excess of 200,000 tons and averaging 10 to 15 ounces in silver.

Secretary Lode Mines Co.

The Secretary Lode Mines Co. owns five claims in the Candelaria District between the Mount Diablo and Lucky Hill mines. In the fall of 1936 this mine was reopened under the direction of

Candelaria, looking east, in the early 1880's was a wooden town of saloons, newspapers, a bank, stores, and stage office. Beyond the town in the foothills is the cemetery; to the right is the road leading to the famed silver mines of Pickhandle gulch, one mile distant (below). The shaft house of the Mt. Diablo mine is shown.

Mark G. Bradshaw of Tonopah. A carload of ore was shipped from the property in September 1936.

The property is developed by a 600-foot shaft and tunnels, which with other workings total about 1 1/2 miles. Equipment includes two Chicago pneumatic compressors and rock drills.

Turquoise and Variscite

Turquoise and variscite deposits were discovered in the Candelaria Mountains in 1908 by A. L. Dees and Edward Murphy. One deposit is several miles northwest of the deserted camp of Columbus and the other is 2 miles west of Rock Hill siding on the Southern Pacific R. R. between Redlich and Coaldale. These deposits have been worked intermittently for gem material when market conditions were favorable.

The production of gem material from these deposits is not a matter of record. According to Carl Reik, who until recently held turquoise claims near Columbus, more than 1,000 pounds of turquoise has been produced by him since 1916.

The Reik group of three claims was sold in 1936 to W. F. Godber, owner of the Western Gem and Jewel Co., 1639 Wooster St. Los Angeles, Calif., wholesale dealers in turquoise. This company uses approximately 25 pounds of turquoise per day for gem stones.

According to Godber, Nevada turquoise is the finest produced in the United States, and much of the material is sold in foreign countries, including England and India, for semiprecious gems.

The turquoise occurs in limestone and shale formation, principally as veinlets along joints or fissures. The veinlets range from knife-blade thickness to a maximum of 1/2 inch. The joints or fissures are quite local and can be traced only a few feet in any direction. The turquoise is closely associated with variscite, which is sometimes mistaken for turquoise.

In October 1936, three men were employed in mining the turquoise. Considerable patience is required because explosives cannot be used and the ground is fairly hard. According to Godber, the quality of the turquoise improves with the hardness of the enclosing rocks. Mining is done mainly in open-cuts, and three men can produce about 1 pound per day.

DOUBLE SPRINGS MARSH DISTRICT

Double Springs Marsh is about 8 miles east of Schurz, a station on the Mina Hazen branch of the Southern Pacific R. R. at the north end of Walker Lake.

The only mining activity on the marsh occurred about 1898, when the Occidental Alkali Co. produced a considerable amount of high-grade soda.

Double Springs Marsh is a typical dry-lake deposit formed by the evaporation of mineral-bearing waters derived from drainage from the surrounding mountains and probably to some extent from hot springs and water of volcanic origin. The dry lake is elliptical in outline, having a length of 4 miles and a width of 1 mile. According to Knapp,[8] the deposit comprises about 800 acres, 500 of which are covered with a deposit of salts ranging from 2 to 14 inches in thickness, averaging about 6 inches. The average composition of the surface salts is as follows:

	Percent
Sodium carbonate	20
Sodium bicarbonate	25
Sodium sulphate	15
Sodium chloride	10
Water	15
Sand and insoluble matter	15
	100

Beneath this surface incrustation is a body of soda clay filled with soda crystals; strong soda solutions constantly rise to the surface by capillary attraction and the salts are deposited by evaporation of the water. This process is slow when the top incrustation is undisturbed, but when that is removed the accumulation of salts on the stripped portion is about 1 inch per year; hence the deposit, when worked, is constantly renewing itself.

The process employed by the Occidental Alkali Co. to produce sodium carbonate has been described by Knapp, and the following has been abstracted from his description: The crude soda stripped from the surface accumulations of salts was first dissolved in hot water in a tank agitated with a revolving arm. When the density of the solution reached about 28° B., the contents of the agitator were drawn into a covered settling tank, where the sand and silt were settled out. The clear solution was drawn off and conveyed by hose to a carbonating cylinder 18 feet long and 6 feet in diameter mounted on rollers and fitted with a 12-inch opening in the center. When the cylinder had been filled the door was sealed and connection made with a carbonic acid receiver, where the gas was stored under pressure of 30 pounds per square inch. When virtually all the carbonates were changed to bicarbonates, the cylinder was discharged, and the solution and the precipitated bicarbonate were sluiced to a cooling and precipitating vat. As the solution cooled, the bicarbonate in suspension was thrown down rapidly when the temperature fell under 70° F. The solution containing chloride and sulphate and some carbonates was drawn to waste leaving the precipitated bicarbonate in the vat. The bicarbonate was washed with cold water to dissolve any sulphate and chloride attached to the crystals of bicarbonate.

After being washed, the bicarbonate was drawn from the vat to a centrifuge, which threw out the remaining solution and left a mass of pure bicarbonate crystals. The bicarbonate was then placed in a stationary, gas-tight retort furnace equipped with stirring arms and externally fired. The gas evolved was conducted through a pipe to a cooling scrubber and thence to a compressor, which compressed and stored it in a receiver. The gas was drawn off from the receiver for use in carbonating a fresh lot of solution. When the calcination was completed, the soda was delivered to a grinding mill, where it was ground and packed for shipment. The finished product was extremely pure anhydrous carbonate.

EAGLEVILLE DISTRICT

Eagleville, also known as the Hot Spring district, is in northeastern Mineral County 11 miles, by road, east of Rawhide. A number of gold-silver deposits in this area have been worked intermittently since the eighties, but only small amounts of shipping-ore have been produced. Principal production, from the viewpoint of tonnage, has been barite from the Highland group of claims.

Highland Group

The Highland group of seven unpatented claims is owned by A. Blundell, of Sparks, Nev., and associates. In 1929 and 1930 this property was leased to the American Development Co. of San Francisco, which shipped 9,000 tons of barite to Oakland, Calif. The barite was mined on a royalty basis of $1 per ton.

Small shipments of gold-silver ore have been mined from veins at the south end of the Highland group.

Development includes three shafts and two tunnels. The deepest shaft is 200 feet. Total workings comprise about 3,500 feet.

The barite occurs in a fissure vein ranging in width from 6 inches to 8 feet. The vein is traceable on the surface for 2,700 feet. The barite is reported to average 96 percent barium sulphate. With favorable market conditions this property can produce a considerable tonnage of barite.

The haul to Fallon, the nearest shipping point, is 50 miles; 32 miles is over macadam highway and 18 miles over fair desert road.

Other Claims

The Rovada Mining Co., owned by Robert Ringling, comprises a group of 12 claims 4 miles south of Rawhide Hot Springs. In 1936 a lessee made four small shipments of gold ore from this property.

At Camp Sunnyside 12 miles northeast of Rawhide Hot Springs is a group of 11 claims owned by Tom Kenyon. Occasional shipments of gold and silver ore have been made from the property by Kenyon.

Mrs. M. R. Wedell of Rawhide Hot Springs owns six scattered claims in this area. Five of the claims have been located for gold and silver and one for molybdenum. All of these claims are in the prospect stage.

FITTING DISTRICT

The Fitting district comprises an area roughly 16 miles long and 6 miles wide in the southeast end of the Gillis Range in central Mineral County. No important mines have been found in this area, and in recent years the metal production has been small. The most recent mining development in this region is the production of andalusite rock from an unusual deposit on the southern flank of the Gillis Range. This is the only commercial occurrence of andalusite in Nevada.

Dover Group

The Dover group of three unpatented claims owned by B. H. Donnelly of Hawthorne is 5 miles east of Thorne, the nearest shipping point. The property is known as the Donnelly andalusite mine.

Andalusite rock was discovered by Donnelly in this area in 1929. In 1936, the property was leased to the Tillotson Clay Co. of Los Angeles for a 3-year period. Six men are employed at the property, and the production up to October 1936 was 450 tons of andalusite rock. The material is shipped to the Tillotson Clay Co. plant at Los Angeles for the manufacture of refractories.

Development work includes a 1 1/2-compartment vertical shaft 50 feet deep, two open-cuts, and a number of surface trenches. Workings comprise a total of 200 feet. The largest open-cut at the west end of the deposit is 30 feet long, 18 feet wide, and 15 feet deep. The shaft has been sunk near this cut, and a crosscut 56 feet long has been driven from the bottom of this shaft. Both shaft and crosscut are in andalusite rock. About 2,500 feet to the east of the shaft the second open-cut has exposed the andalusite for a width of 25 feet, a length of 20 feet, and a depth of 12 feet.

Equipment on the property includes an Ingersoll Rand Imperial Type 14 portable compressor, an Essex geared hoist driven by a gasoline engine, and a blacksmith shop. Hoisting is done with 1,000-pound capacity bucket.

The andalusite formation strikes approximately east and west. Near the shaft the andalusite rock is covered with a mantle of surface debris and clay up to 12 feet thick. The clay and surface debris are traversed with seams of gypsum.

In addition to andalusite, the rock contains corundum, quartz, sericite, and probably a little dumortierite. There is considerable variation of the mineral constitutents in the rock. Corundum predominates in some specimens, while in others andalusite is the most abundant mineral. After the material is mined it is hand-sorted into two grades, according to specific gravity and mineral constitutents.

Insufficient work has been done to block out tonnage, but in all probability this deposit contains large reserves.

Other Andalusite Claims

Five miles east of the Donnelly property is a group of claims located for andalusite. This group is owned by Joseph Maltesta of Hawthorne and associates.

Another group of 5 claims near the Donnelly property is owned by Peter Vuich and William Ray of Hawthorne. This group of claims is also under lease to the Tillotson Clay Co.

Only a small amount of development work has been done on either of the aforementioned properties, and no shipments have been made.

Chiatovich Group

A deposit of bentonite occurs about 9 1/2 miles north of Hawthorne in the foothills of the Gillis Range near the south end of Walker Lake. The nearest shipping point is Thorne, 3.3 miles south.

The deposit is covered by three unpatented claims owned by Martin M. Chiatovich, of Hawthorne. In 1936, the deposit was worked on a royalty basis by the Naval Ammunition Depot at Hawthorne. The bentonite used to prevent water seepage in a reservoir constructed at the Depot.[9] Up to October 1936, the Navy had used approximately 1,500 tons for this purpose. The bentonite is mined by hand shoveling.

The bentonite outcrops in places over an area at least 2,000 feet long and several hundred feet wide. Portions of the deposit are covered with iron-stained soil to a depth of 3 feet or more. No attempt has been made to determine the extent of the deposit, but from surface indications a large tonnage is indicated.

Mica

On the north slope of the Gillis Range 22 miles northeast of Hawthorne via Ryan Canyon is an occurrence of mica. No production has ever been made.

Mica of the muscovite variety is exposed in a shallow shaft sunk 12 feet on the deposit. The mica is in three parallel veins that dip about 45°. The veins range in width from 12 to 20 inches. Country rock is granite. The mica is present in the veins as an intergrowth of crystals that average several inches in area, and in blocks up to 1 inch thick. Its color in thick plates is dark green, but in thin sheets it appears colorless and transparent. Most of the plates show checks and corrugations.

The material occurs in sheets too small for sheet mica. It is probably suitable only for grinding purposes.

Graphite

On the west side of Winwan Flat, about 26 miles south of Rawhide and in the vicinity of the mica deposit, is a deposit of graphite. It was discovered by J. A. Kelly, of Rawhide, in 1916. None has ever been produced.

The graphite occurs as a vein in granite. The vein has been prospected by several open-cuts.

The workings were caved so that the character of the material and the probable extent of the deposit could not be determined.

Iron

Deposits of iron ore are reported to occur in the southern part of the Gillis Range about 10 miles northwest of Acme Siding on the Southern Pacific R. R. The deposits are inaccessible by automobile. Several prospectors have informed the writer that the deposits are quite extensive and consist principally of hematite.

Fick Mining Company

The Fick Mining Co. property, controlled by H. J. Fick of Hawthorne, comprises 29 unpatented claims in four groups. The Gold Basin, Greenman, and Paymaster groups are in the Fitting district several miles northeast of Kinkaid Siding, and the 1917 group is in the Hawthorne district. This company was incorporated about 1926 as a reorganization of the Nevada Ore and Copper Co. In recent years the property has been worked sporadically by lessees, who produced only a little.

In 1936, J. R. Parker and associates acquired the Fick Mining property under bond and lease, and several men were employed in rehabilitating the mill preparatory to retreating the tailings dump.

It is said that the tailings, comprising about 8,000 tons, average $5.40 in gold. These tailings were derived from ore taken from the La Panta, Montreal, and other mines in this area. Parker planned to install flotation equipment in the mill to recover the values in the tailings.

The mill, situated a short distance east of Kinkaid Siding, is equipped with a small crusher, a 4- by 4-foot ball mill, and a Wilfley table. Water for milling is available from a well 90 feet deep hear the millsite.

The Fick Mining Co. properties are developed by several shafts, the deepest of which is 600 feet, and over 1 mile of underground workings.

Formations are principally shale and limestone intruded by quartz monzonite. Ore occurs as lenses and stringers along bedding planes of the limestone. Values are in gold, silver, and copper.

Hawaiian Group

The Hawaiian group of four unpatented claims owned by B. H. Donnelly, of Hawthorne, is in Ryan Canyon 5 1/2 miles northeast of Thorne, a station on the Hazen Mina branch of the Southern Pacific R.R. This property was discovered in 1906 by a man named Ryan, after whom the canyon was named.

No production has ever been made from the property. Prospecting work consists of a shaft 50 feet deep and some surface cuts, evidently made in search for shipping ore which was not found in commercial quantities. Property has been idle for many years.

The nearest water for milling is on the flat near Thorne, approximately 1,500 feet lower than the deposit.

The interesting feature of the property is a prominent outcrop ranging from 150 to 600 feet in width and traceable on the surface for nearly a thousand feet. This outcrop shows considerable silicification, and, according to Donnelly, samples taken at various places along the outcrop assayed $2.50 to $14 per ton. Values are in gold and silver associated with small amounts of copper, manganese, and iron oxides. Sampling done has been insufficient to determine the average value of the material.

GARFIELD DISTRICT

The Garfield district is at the northern end of an unnamed mountain range situated south and west of Soda Spring Valley. Acme, a siding on the Southern Pacific R. R., is 6 miles north. The district is accessible by automobile road either from Mina, 20 miles southeast, or Hawthorne, 24 miles northwest.

Silver-gold ore was first discovered in this area by Joshua Mass and Amos Everson in 1882. From 1882 to 1887, the Garfield mine (formerly known as the Blue Light) is said to have produced several million dollars in shipping ore. About 1890 an English company called the Hampton Plain Exploration Co. acquired the Garfield mine and erected a 10-stamp mill at Garfield Springs 9 miles south of the mine. To judge from the tailings at Garfield Springs, roughly 5,000 tons of ore were treated. In 1935, an unsuccessful attempt was made to treat these tailings by cyanidation.

About 1922 the Mabel mine, adjoining the Garfield, was acquired by the West End Consolidated Mining Co. Since 1922 this mine has been a small but consistent producer of high-grade shipping ore. From 1922 to 1929 the Mabel mine is reported to have yielded 4,310 tons of ore having a gross value of $421,627, an average of $97.83 per ton.

In recent years mining activity in the Garfield district has been confined to small leasing operations.

Mining equipment includes a Chicago Pneumatic two-drill compressor and a Fairbanks-Morse gear hoist, both powered by gasoline engines. Hoisting is done with bucket and crosshead.

In 1936 that portion of the mine above the 600-foot level was under lease to Lloyd Wilson, of Hawthorne, and associates. The lessees had produced about $18,000 of shipping ore in about one year. The royalty payments based on the gross value of the ore are 20 percent on ore having a value of less than $100 per ton and 25 percent over $100 per ton. The haul to Mina, a distance of 23 miles, costs $2.75 per ton on contract.

Veins range in width from 3 inches to 3 feet and contain silver, gold, and lead.

HAWTHORNE DISTRICT

The Hawthorne district includes a large area tributary to the town of Hawthorne, which is at the south end of Walker Lane 7 miles southwest of Thorne, a station on the Mina-Hazen branch of the Southern Pacific R.R. It is sometimes divided into the Lucky Boy section and the Pamlico section, 6 and 10 miles, respectively, southeast of Hawthorne.

The Pamlico and La Panta mines in the Pamlico Range were the most important early-day producers. Other early-day mines of less importance in the Pamlico section were the Good Hope, War Eagle, New York Central, and Gold Bug. In later years, the Lucky Boy has been the principal property in the Hawthorne area. In recent years there has been little mining activity.

Production of the Hawthorne District from 1910 to 1934 is shown in table 4.

Lucky Boy Consolidated Mines Co

The Lucky Boy Consolidated Mines Co., J. H. Miller, of Hawthorne, president and principal owner, owns 14 patented claims on the east slope of the Wassuk Range. The Lucky Boy mine was discovered in 1906 by Guy E. Pritchard while working on the road over the Lucky Boy Pass.

In 1936 the property was under option to eastern interests, and several men were employed in cleaning out the Miller tunnel. This work was done primarily to hold the option.

I.C. 6941
TABLE 4. - Gold, silver, copper, and lead production from Hawthorne District, Mineral County, Nev., 1904-35, in terms of recovered metals

(Compiled by Charles White Merrill, Mineral Production and Economics Division, U. S. Bureau of Mines)

Year	No. of mines	Placer Gold Fine oz.	Placer Gold Value	Placer Silver Fine oz.	Placer Silver Value	Total value	No. of mines	Ore, short tons	Lode Gold Fine oz.	Lode Gold Value	Lode Silver Fine oz.	Lode Silver Value
1904	---	---	---	---	---	---	1	1,900	142.99	$2,956	35	$20
1905	---	---	---	---	---	---	---	---	---	---	---	---
1906	---	---	---	---	---	---	---	---	---	---	---	---
1907	---	---	---	---	---	---	2	35	25.54	528	9	6
1908	1	47.99	$992	7	$4	$996	1	64	45.96	950	8	4
1909	---	---	---	---	---	---	4	1,815	481.62	9,956	408	212
1910	---	---	---	---	---	---	15	4,090	864.05	17,861	421,384	227,547
1911	---	---	---	---	---	---	9	1,323	152.68	3,156	100,859	53,455
1912	---	---	---	---	---	---	15	1,358	177.86	3,677	120,854	74,325
1913	---	---	---	---	---	---	9	452	115.87	2,395	1,132	684
1914	---	---	---	---	---	---	8	305	219.95	4,547	1,098	607
1915	2	70.13	1,450	15	8	1,458	9	405	238.83	4,937	13,211	6,698
1916	2	23.31	482	4	3	485	8	377	27.09	560	56,439	37,137
1917	2	3.50	72	2	1	73	15	535	66.54	1,376	17,445	14,375
1918	---	---	---	---	---	---	8	105	41.28	853	10,019	10,019
1919	---	---	---	---	---	---	8	59	11.50	238	3,211	3,596
1920	---	---	---	---	---	---	2	256	43.00	889	15	16
1921	---	---	---	---	---	---	2	22	23.29	481	9	9
1922	---	---	---	---	---	---	6	102	31.14	644	11,110	11,110
1923	---	---	---	---	---	---	8	99	250.63	5,181	2,237	1,834
1924	---	---	---	---	---	---	8	46	151.82	3,138	200	134
1925	---	---	---	---	---	---	4	31	68.89	1,424	3,311	2,298
1926	---	---	---	---	---	---	3	18,549	27.10	560	103,473	64,567
1927	---	---	---	---	---	---	4	219	17.30	358	17,424	9,879
1928	---	---	---	---	---	---	3	210	9.32	192	11,854	6,935
1929	---	---	---	---	---	---	4	104	36.48	754	471	251
1930	(2)	(2)	(2)	(2)	(2)	(2)	5	(2)	(2)	(2)	(2)	(2)
1931	---	---	---	---	---	---	(2)	(2)	(2)	(2)	(2)	(2)
1932	---	---	---	---	---	---	3	52	84.40	1,745	1,538	434
1933	(2)	(2)	(2)	(2)	(2)	(2)	5	84	176.98	4,524	148	52
1934	(2)	(2)	(2)	(2)	(2)	(2)	8	319	460.70	16,101	256	165
1935	(2)	(2)	(2)	(2)	(2)	(2)	9	432	697.62	24,417	11,507	8,271
Total 2/		154.52	$3,331	30	$17	$3,348		33,412	4,700.43	$114,605	909,710	$534,646

2/ Bureau not at liberty to publish figures but disclosed figures included in totals. Continued ---

TABLE 4. - Gold, silver, copper, and lead production from Hawthorne District, Mineral County, Nev., 1904-35, in terms of recovered metals (Continued)

(Compiled by Charles White Merrill, Mineral Production and Economics Division, U. S. Bureau of Mines)

Lode (Continued)

Year	Copper Pounds	Copper Value	Lead Pounds	Lead Value	Total value	Average recoverable value of ore per ton 1/	Total value, lode and placer
1904	---	---	---	---	$2,976	$1.57	$2,976
1905	---	---	---	---	---	---	---
1906	---	---	---	---	534	15.26	534
1907	---	---	---	---	954	14.91	1,950
1908	---	---	---	---	10,168	5.60	10,168
1909	42,966	$5,457	---	---	280,427	68.56	280,427
1910	1,709	214	671,860	$29,562	62,529	47.26	62,529
1911	52,317	8,632	126,787	5,704	94,972	69.94	94,972
1912	38,131	5,910	185,284	8,338	8,989	19.89	8,989
1913	5	1	---	---	5,396	17.69	5,396
1914	1,594	279	6,169	241	12,586	31.08	14,044
1915	18,537	4,560	14,293	672	47,087	124.89	47,572
1916	35,558	9,707	70,000	4,830	27,937	52.22	28,010
1917	474	117	28,828	2,479	12,309	117.23	12,309
1918	3,133	583	18,592	1,320	4,556	77.22	4,556
1919	---	---	2,631	139	905	3.54	905
1920	---	---	---	---	490	22.27	490
1921	38	5	17,681	972	12,731	124.81	12,731
1922	263	39	6,557	459	7,513	75.89	7,513
1923	1,604	210	3,330	266	3,748	81.48	3,748
1924	---	---	5,750	500	4,222	136.19	4,222
1925	7,958	1,114	142,195	11,375	77,616	4.18	77,616
1926	1,212	159	21,978	1,385	11,781	53.79	11,781
1927	802	115	18,106	1,050	8,292	39.49	8,292
1928	5,888	1,036	---	---	2,041	19.62	2,041
1929	(2)	(2)	(2)	(2)	(2)	(2)	(2)
1930	(2)	(2)	(2)	(2)	(2)	(2)	(2)
1931	119	7	12,086	375	2,561	49.25	2,561
1932	---	---	---	---	4,576	54.48	4,576
1933	310	25	750	28	16,319	51.16	16,319
1934	1,386	115	8,088	324	33,127	76.68	33,127
Total 2/	214,004	$38,285	1,362,550	$70,098	$757,634	$22.68	$760,982

1/ Not to be confused with average assay value of ore. 2/ Bureau not at liberty to publish figures but; disclosed figures included in totals.

The Lucky Boy mine is developed by the Hubbard two-compartment shaft 950 feet deep, inclined 70°, the Miller tunnel 6,400 feet long, and approximately 2 miles of subsidiary workings.

Mining equipment includes a Sullivan compressor, 50-horsepower electric hoist, electric haulage locomotive, and other mining machinery.

A 125-ton flotation mill was erected in 1926 near the portal of the Miller tunnel. Milling equipment includes a 10- by 12-inch Blake crusher, 2 1/2- by 8-foot Hardinge ball mill, Dorr Duplex classifier, three K & K flotation units, a Deister table, and an Oliver filter.

Electric power is available at the property.

Water for milling is available from Summit Springs and from the mine. Water from the springs is brought to the millsite by gravity pipeline 4 miles long and 3 to 4 inches in diameter.

According to Hill[10], the formations at the Lucky Boy mine are granodiorite intrusive into limestone. The ore deposit is a complex contact vein. The ore occurs as lenses and shoots with a steep west pitch. The main vein occurs in an irregular fracture. The fracture took place after the intrusion and cuts the limestone, granodiorite, and contact, although in general it follows the contact rather closely. The width of the vein varies from 1 to 8 feet and has a dip of from 70° to 80°. The ore minerals are argentiferous tetrahedrite, galena, argentite, horn silver, and a little sphalerite and gold.

La Panta Mine

The La Panta property comprises four claims controlled by E. H. Daugherty of Carson City, Nev. It is 12 miles southeast of Hawthorne, and is reported to have produced about $300,000, principally in gold. In 1936 the property was idle.

Development work includes a number of shafts, the deepest of which is 300 feet. Total workings comprise about 5,000 feet. Some of the mine dumps have been screened and shipped in former years. All equipment has been removed from the property.

The ore deposits are irregular replacement bodies in limestone near a basic dike, probably basalt. The values are mainly free gold in a gangue composed principally of iron oxides. Several samples panned by the writer carried values estimated at $6 to $10 per ton.

Pamlico Mine

The Pamlico mine 10 miles southeast of Hawthorne, comprises eight claims controlled by E. H. Daugherty of Carson City, Nev.

In the seventies and eighties this mine is reported to have produce& several hundred thousand dollars. In later years the mine has been worked intermittently by various lessees. In 1936 the property was idle.

The Pamlico vein is developed by shafts, drifts, raises, and crosscuts aggregating several miles in length.

Equipment includes the remains of a 20-stamp mill. All the mill equipment has been removed, with the exception of the stamps (1,050 pounds each) and a Gates No. 3 gyratory crusher.

In former years water for milling was piped from Cottonwood Creek, which heads near Buller Mountain in the Wassuk Range, 10 miles southwest of the mine.

The tailings pile below the mill is roughly estimated to contain 1,500 tons. These tailings have been sampled but their average value could not be learned.

The Pamlico vein is said to be in rhyolite. The ore consists of iron-stained quartz that carries free gold as nuggets and wires and some argentiferous galena.

Placer Gold

Placer gold has been found in the Canyon below the Pamlico mine. About $8,000 in placer gold is said to have been mined here about 1912. The gravels are deep; one shaft was sunk 170 feet before bedrock was encountered.

Barite

Barite occurs in a range of low hills 13 miles east of Hawthorne. Considerable barite was shipped from this locality from 1916 to 1919 to consumers on the Pacific coast. Most of the production was made from the Crystal claim, patented in 1926, by the D. V. 0. Products, Inc. The nearest shipping point is Kinkaid Siding, 1.3 miles from the property.

Development work on the Crystal claims includes three shallow shafts and several hundred feet of lateral workings. In 1936, the property was idle and all mining equipment except a small geared gasoline hoist and 1-ton skip had been removed.

The barite is in a nearly vertical vein. On the surface the vein is exposed by trenches and open-cuts for a length of about 800 feet. The width of the barite varies from 1/2 to 8 feet. At the northerly end of the vein the barite is associated with copper minerals.

KING DISTRICT

The King mining district is in northeast Mineral County 14 miles east of Rawhide and west of Mount Anna, a prominent landmark. It is accessible by automobile either from Fallon 50 miles to the northwest or from Hawthorne via Ryan Canyon about the same distance to the south. This district was prospected to some extent many years ago and abandoned until 1926, when a small stringer of rich gold ore was found at a depth of about 10 feet in an old shaft. This strike attracted attention, and in 1926 and 1927 up to 100 men prospected in the area. The only production was one carload of shipping ore reported to average about $40 per ton in gold.

Donnelly Group

The principal property is the Donnelly group of nine claims and two fractions of unpatented ground owned by B. H. Donnelly and R. L. Mundell of Hawthorne, Nev. Total development work on this ground is about 800 feet and consists of a crosscut tunnel 550 feet long, a shaft 100 feet deep, and several shallower shafts and a number of open-cuts along the course of the mineralization. Most of this work was done by lessees in searching for shipping ore.

The formations are principally andesite and rhyolite intruded into granite. The mineralization appears to be associated with a dike 40 to 50 feet wide. The mineralized zone is roughly 4,000 feet long and 60 to 200 feet wide, traversed with numerous seams and small fissures filled with quartz carrying gold, silver, lead, pyrite, and a little copper. The individual seams are too small and irregular to be mined separately, but thorough sampling may show that the seams are numerous and rich enough to carry the intervening rock making a low-grade ore.

Sporadic sampling in places by Mr. Donnelly showed values varying from $1.60 to $16 per ton. In addition, numerous pannings indicated a wide distribution of values. More thorough sampling is necessary to determine the value of the deposit. Probably the only hope for the district is the possibility of developing a large tonnage of milling grade ore.

MARIETTA DISTRICT

The Marietta, also known as the Black Mountain district, is in the vicinity of the old camp of

Marietta 25 miles by automobile road southwest of Mina and 10 miles west of Belleville, the latter a station on the narrow-gage railroad that runs between Mina and Keeler, Calif.

The Endowment mine was discovered in this area in the early sixties, shortly after the discovery of Aurora. In recent years the principal property has been the Moho mine.

Moho Mine

The Moho mine, on the south slope of the Excelsior Mountains, is accessible by automobile road from Mina 15 miles distant. The mine was located in 1903, and considerable work was done in searching for shipping ore. Property is credited with a production of $75,000, made mostly by lessees.

With the increase in the price for gold, interest in the property was renewed. It was sampled by Earl Bohannon; and a small company, called the Federal Mining and Engineering Co., acquired 12 unpatented claims. In 1935, a pilot milling plant employing cyanidation was erected. This plant has a capacity of 25 tons per day, and up to October 1936 about 1,500 tons of dump ore had been milled.

The mill equipment includes a Dorr classifier in closed circuit with a home-made rod mill 5 feet long and 28 inches in diameter, 2 Parral agitators, a clarifier, and 2 solution tanks. Power for milling is furnished by a 60- horsepower Holt gasoline engine.

Dump material is fine, so that primary crushing is not necessary. Ore is ground to minus 60-mesh in a 2-pound-strength cyanide solution. Lime consumption is 7 pounds and cyanide consumption 1/2 pound per ton. Precipitation is effected with zinc shavings. The recovery by cyanidation is reported to be 90 percent on ore averaging about $8 per ton.

Water for milling is pumped from Marietta Springs through a 2-inch- diameter pipe line 4,500 feet long. Approximately 25 gallons of water per minute are pumped by Gould Triplex pump, size 3 1/2 by 5 inches, driven by Fairbanks-Morse Z-type, 10-horsepower, gasoline engine. Water is under head of 350 feet.

Mine development consists of one tunnel driven 1,200 feet on a vein, nine shorter tunnels, and several shafts, the deepest of which is 300 feet. Total underground workings comprise 3,500 feet.

The formation is mainly andesite. The ore occurs in a series of veins that average 3 1/2 to 4 feet in width and dip about 70°. One vein is traceable on the surface for a distance of 1 mile. The values are chiefly in gold, with some silver. Lead in the form of cerussite is present in the ore in amounts. averaging about 2 percent.

Before the property was purchased it was sampled by 900 cut samples and, according to Bohannon, at least 100,000 tons of ore averaging $9 per ton in gold and silver have been blocked out. During the last 2 years several sets of lessors working on the property produced $12,000 in shipping ore. Mining is done by hand. The royalty payments are 15 percent of the net smelter returns on ore having a value of $30 per ton or less and 20 percent above $30 per ton.

The smelter returns on a shipment of ore made by W. J. Forbes on March 1, 1935, to American Smelting & Refining Go., furnished the following data:

Metal Quotations:
    Au $34.9125 per oz.
    Ag   .64 1/8 per oz.

Settlement assay:
    Au   1.015 oz.
    Ag   6.3 oz.
    Pb   1.4 percent

Value	Au at $31.81825 per oz.	$32.30	
	Ag less 0.5 oz. at $0.64125	3.72	
	Total	36.02 per ton	

Treatment charge:
    Base charge $3.50
    10% excess over $20   1.60
                                   5.10
    Labor increase   .06
                                   5.16 per ton

	Pounds		
Wet weight	119,164		
Less 10.48% moisture	12,488		
Net weight	106,676	53.338 tons at $30.86	$1,646.01
Freight advanced at $4.30 per ton		$245.87	
Demurrage		6.00	
Fed. Min. & Engr. Co., 15 percent		207.57	
			459.44
Net proceeds			1,186.57

The trucking cost to railroad is $1.50 per ton; distance of haul, 13 miles.

Endowment Mine

The Endowment mine is 3 miles northerly from the old camp of Marietta. This property comprises five unpatented claims owned by B. F. Baker and A. V. Reeves of Mina, Nev.

Property was idle from 1884 until 1923, when it was worked by lessees. In 1926 lessees shipped 144 tons of ore averaging 81.6 ounces in silver, $2.34 in gold, 3.83 percent copper, 5.25 percent lead, and 7.6 percent zinc.

Production from this mine, according to statements of men familiar with the property, has been approximately $150,000, although some estimates in earlier reports give a figure as high as $1,500,000 prior to 1884.

In recent years the property has been idle.

Development consists of a tunnel reported to be 500 feet long, a winze 333 feet deep sunk from the tunnel level, and several other winzes.

Two veins occur in quartzite and quartzite conglomerate with inter-bedded argillite. The width of the veins ranges from several inches to 6 feet. The principal values are in silver associated with cerussite, smithsonite and copper carbonates near the surface, while at depth galena, sphalerite, pyrite, and a little chalcopyrite are associated with silver.

The following account of the mine taken from an early report by Whitehill[11] is of interest.

> The vein matter is decomposed and carries carbonate of lead, argentiferous galena, and iron. By mill process the ore yields from $50 to $125 in silver. A tunnel has been run in on the vein a distance of 500 feet and connects with a shaft at a depth of 200 feet. From this tunnel level two winzes have been sunk 200 feet, a distance of 300 feet apart. The pay ore is about 5 feet in width, though the vein matter is much wider. The 5-stamp quartz-mill, erected at Philadelphia by the State of Nevada for the Centennial Exhibition, was removed to this district and has run very successfully on the ores obtained here. The Endowment is an incorporated company, the stock of which is listed in the San Francisco Stock and Exchange Board. The assessments levied have been $50,000 and the yield of bullion $88,764.

Rutty Group

The Rutty group of three unpatented claims owned by Joe Rutty of Marietta is on the south slope of the Excelsior Mountains about 1 1/2 miles north of the camp. Rutty, an old time Cripple Creek miner, has worked his property by himself since 1910 and made a living from it.

The mine workings comprise four adits driven at various elevations, which, with subsidiary workings, total about 4,000 feet. The ore is mined by hand and is packed down the hill on burros for trucking to the mill.

The Rutty mill is a small affair equipped with a 12- by 8-inch jaw crusher, two stamps weighing 1,050 pounds each, and an amalgamation plate 8 feet long and 3 feet wide. The stamps and crusher are driven by an old automobile engine. Water for milling is pumped by a 3-horsepower Fairbanks-Morse Z-type engine from a well 23 feet deep near the mill site.

The ore bodies are small and bunchy and occur along cross fractures in limestone. Values are mainly in gold. According to Rutty considerable ore that will average $6 per ton at current metal prices is developed on his property. This ore is too low grade to be mined under existing conditions.

Gold Gulch Mining & Milling Co.

The Gold Gulch Mining & Milling Co. was incorporated in 1929 as a reorganization of the London Silver Lead Mines Co. In December 1934 this property was sold at sheriff's sale to C. E. Flagg of Reno, Nev. The property which has been idle for several years, comprises 14 claims several miles northeast of Marietta and about 12 miles from Belleville.

Development work consists of a two-compartment shaft 150 feet long and some lateral workings. In 1928 a mill was erected at the Marietta. Mill equipment includes a Dodge crusher, a Denver quartz mill, and two concentrating tables powered by a Fairbanks-Morse gasoline engine.

Judging from the size of the tailings dump, the mill treated only a few tons of ore.

Annett Group

Several miles east of Marietta is a barite deposit covered by four un- patented claims owned by Al Annett of Mina. Belleville, on the Mina-Keeler narrow-gage railroad 11 miles east of the property, is the nearest shipping point. No production has ever been made.

Development consists of a tunnel 80 feet long and several open-cuts. The barite is in a series of parallel veins, each averaging about 6 feet in width.

MOUNTAIN VIEW DISTRICT

The Mountain View or Granite district is near two abandoned camps of the same name, about 1 1/2 miles apart, near the crest of the north end of the Wassuk Range 9 miles northwest of Schurz.

Gold was discovered at Mountain View by William Wilson in 1904, when the area formed part of the Walker Lake Indian reservation. In 1906, part of the reservation was thrown open to prospectors, and several claims were located at Mountain View by William Wilson. About 1908 some gold ore carry- ing some silver was mined and treated at a 5-stamp mill then operating at Nordyke, 15 miles from Mountain View. Later, a small stamp mill operated for a short time at Mountain View.

Small amounts of copper ore carrying a little gold and silver and some oxidized gold ore have been shipped by lessees, but in recent years there has been no activity in the area.

The formation is principally granodiorite. At the Mountain View mine the mineralization follows a narrow decomposed diorite dike that dips about vertical. The vein filling is principally quartz and iron oxide.

A vein 8 to 25 feet in width at the surface shows copper carbonate in a gangue of quartz and crushed granodiorite. A tunnel several hundred feet long was driven to prospect this deposit at depth, but no commercial ore was found.

MOUNTAIN GRANT DISTRICT

The Mountain Grant, also known as the East Walker district, covers that portion of the Wassuk Range that lies between Cory Canyon on the south and Cottonwood Canyon on the north. A considerable portion of this area has been withdrawn from the public land in order to establish the Naval Ammunition Depot at Hawthorne.

The area in the vicinity of Mountain Grant is accessible by automobile road via either Cottonwood or Cory Canyon. This road was constructed in 1934 and 1935 by the Civil Conservation Corp under the supervision of the Forest Service and the Naval Ammunition Depot at Hawthorne.

The principal property in the district in the early days was the Big Indian mine. In recent years the only mining activity has been small-scale operations by owners or lessees. For the past 3 years the principal property has been a small-scale placer operation in Laphan Meadows.

Grant Mountain Gold Mine

The Grant Mountain Gold Mine is owned by C. B. Murray, of Reno, and associates. This company owns 13 unpatented placer claims at Laphan Meadows, several miles southwest of Mountain Grant and 24 miles by road west of Hawthorne, Nev.

Placer operations were begun in Laphan Meadows in 1906, when this area, formerly part of the Schurz Indian reservation, was thrown open to prospectors. Small-scale placer operations have been carried on intermittently for many years, but no figures are available on the total production of placer gold.

Placer operations have been carried on by the Grant Mountain Gold Mine Co. during the summer months for the past 3 years with a crew of four men. The placer season lasts about 6 months of the year.

Production for the first two seasons amounted to $8,000.

Placer equipment comprises a 3/4-cubic yard P. & H. dragline mounted on caterpillars, and a portable sluice. The sluice is mounted on skids so that it can be moved with the dragline equipment. The hopper of the sluice is 15 feet above the ground. An inclined rail grizzly, 6 by 8 feet in area, with 4-inch openings, is above the hopper. The grizzly oversize is discharged by gravity at the side of the machine. The grizzly undersize drops into a bin, and from the bin the gravel is fed to a sluice 50 feet long, 16 inches wide, and 10 inches deep, sloping 1-inch per foot and equipped with Hungarian riffles.

Water for sluicing is obtained from a number of small springs in the vicinity. About 1 miner's inch of water is available during the summer months. The water is reclaimed from two settling ponds below the sluice. Water from the settling ponds is returned to the head of the sluice with a 5-inch Byron Jackson centrifugal pump belt driven by a 6-cylinder automobile engine.

In mining, about 2 yards of overburden are stripped for each yard of material put through the sluice. The operating costs are about 25 cents per cubic yard of gravel treated.

The gold occurs in a channel that averages about 40 feet in width and 9 feet in depth. The bedrock is decomposed granite, and the best values occur on and in granite. The alluvium is mainly a black loam with no large boulders. The average value of the ground is reported to be 55 cents per cubic yard, including 6 feet of barren black loam.

The gold is fairly coarse and shows abrasion. The fineness averages 898. The largest nugget found in 1935 had a value of $30. A considerable amount of black sand is present in the sluice concentrates. The gold probably is derived from a series of gash veins in granite.

Big Indian Mine

The Big Indian mine is on the north fork of Cory Canyon 29 miles from Hawthorne via Cottonwood Canyon. The property comprises 14 unpatented claims owned by Mrs. Julia Ward of Reno, Nev.

In the early days a six-stamp mill was erected in Cory Canyon to treat ore from this mine. Production is said to have been about $100,000.

Property is developed by several tunnels and a number of shallow shafts. In 1936 the property was idle. There is no equipment on it.

The ore is in a series of fissure veins in granodiorite. Values are chiefly in gold in a quartz gangue.

Cory Mine

The Cory mine is on the south fork of Cory Canyon. Sam Kelso of Hawthorne owns the Cory group of six unpatented claims. In the seventies a small mill was operated at the mouth of Gory Canyon on ore from this mine.

Development work consists of several tunnels and shallow shafts that total approximately 3,000 feet.

Ore occurs in fissure vein in granodiorite. Values are chiefly in silver.

Talisman Group

The Talisman group of 13 claims owned by Ward Daves and Michael Knopf of Hawthorne is in Cottonwood Canyon.

Development work consists of a number of short tunnels.

Ore is in fissure veins in granodiorite. Gold is the principal metal.

One of the claims of the Talisman group contains pumice. The pumicite bed is exposed by a shallow open cut 10 feet long. The pumicite dips about 30° and averages about 3 1/2 feet thick.

Return Group

The Return group of two unpatented claims owned by J. W. Mallory of Hawthorne is on the west slope of Mountain Grant. The only production from this property has been about $200 in shipping ore.

Development work consists of a tunnel 175 feet long.

Ore occurs in a fissure vein in granodiorite. The vein has been disturbed considerably by faulting. Gold occurs in a gangue of quartz.

Molybdenite

According to D. H. Donnelly of Hawthorne, molybdenite is present in granite in Cory

Canyon. In former years the deposit was prospected by several open-cuts.

Molybdenite occurs on ground included in the Naval Ammunition Depot at Hawthorne.

MOUNT MONTGOMERY DISTRICT

Mount Montgomery District is in southwestern Mineral County at the north end of the White Mountain Range. In this area gold, silver, quick silver, and fluorspar ores have been taken from a number of small mines. When the district was visited in October 1936 there was very little mining activity.

Mount Montgomery Quicksilver Co.

The Mount Montgomery Quicksilver Co. comprises four unpatented claims owned by W. H. Kirkbride, 1067 Fifth Ave., New York City. This property is in the White Mountains 5 1/2 miles south of Mount Montgomery station. Cinnabar was discovered here by John C. Morris in 1916. The claims were sold to F. C. Beedle, of Reno, Nev., who produced $16,000 in quicksilver during the war period. In 1936 the property was worked under lease for 4 months by Walter L. Lowe. Lowe produced only 1 flask of quicksilver.

Development includes two tunnels, respectively 300 and 400 feet long and other workings totaling about 1,500 feet. Mining has been done by hand methods. Two D-type retorts are on the ground.

Cinnabar is present in bunches in opalized rhyolite, which is considerably shattered.

Tip Top Mine

The Tip Top mine is about 10 miles south of Mount Montgomery station, which is on the narrow-gage railroad between Mina, Nev., and Keeler, Calif. This mine was worked from 1913 to 1915 by the Atkins-Kroll Co. of San Francisco, which equipped the property with a 10-stamp cyanide mill of 50-ton capacity. About $100,000 in gold and silver is reported to have been produced.

Development comprises several tunnels and other workings that total about 1,000 feet.

The formations are said to be andesite and rhyolite. Two veins have been found on the property.

Golden Gate Mining Co.

The Golden Gate Mining Co. owns seven unpatented claims 4 miles east of Queens and about the same distance west of Mount Montgomery station. The principal owner is A. A. Isaacson. The property was located in 1915 by Al Stevens, of Bishop, and associates. In 1924 and 1925 some high-grade ore was mined and treated in a small mill at Queens. In 1936, two men were carrying on development work. Production, principally of gold, is said to have been about $50,000.

Development consists of several tunnels, one about 1,000 feet long and other workings totaling about 3,500 feet.

Mining equipment includes a Curtis 1-drill compressor driven by a 4-cylinder Climax gasoline engine, rock drills, and a blacksmith shop.

The formations are principally limestone and granite. Gold-silver ore is present in a vein ranging in width from 2 to 7 feet and dipping about 60°. The vein is considerably faulted.

Other Properties

The Shang mine 4 miles west of Mount Montgomery station comprises six claims owned by

Al Stevens of Bishop, who located the property about 1924. Development consists of several shallow shafts and tunnels, totaling in all about 1,000 feet. Formation is limestone and porphyry and values are in gold and silver. Small amounts of shipping ore have been produced.

The Silver Trail group, adjoining the Shang property on the south, is owned by Al Stevens, of Bishop, and associates. Development consists of about 1,000 feet of shallow shafts and tunnels. Formation is limestone and andesite. Values are in silver, lead, and a small amount of gold. At least 7 carloads of ore have been shipped.

The McElroy mine is near the Tip Top mine. This property is said to have been acquired recently by Dr. D. E. Mitchell and associates.

Mogoe Claims

In 1929 cinnabar was discovered by Joseph Mogoe 8 miles northwest of Basalt. Mogoe owns eight claims in this area. Four flasks of quicksilver were produced in 1930 and 1931.

Development consists of several shallow shafts, a tunnel 85 feet in length, and a number of open-cuts, totaling about 400 feet. Mining has been done by hand, and equipment includes a D-type retort.

Cinnabar is present in quartzite and limestone.

Bentonite

One mile west of Mount Montgomery station a clay deposit of bentonite type is covered by one unpatented claim owned by Robert D. Somerville, of Basalt. The highway between Coaldale, Nev., and Benton, Calif., passes over the deposit, and the Mina-Keeler narrow-gage railroad is within 1/2 mile. The only development is a tunnel 40 feet long. No production has been made.

The bentonite is traceable on the surface for about 1,000 feet. The width is approximately 100 feet. No physical tests have been made on the bentonite to determine its value for specific purposes.

Fluorspar

In 1914 George F. Thompson, of Mina, and Paul Kienapfiel (deceased) discovered fluorspar 3 miles south of Mount Montgomery station. Five carloads were shipped from here in 1926. The deposit has been explored by several tunnels totaling 1,000 feet. Fluorite associated with considerable quartz is present in an andesite formation in veins from 1 inch to several feet in width. Four claims have been relocated recently on the deposit by Robert D. Somerville of Basalt.

ONEOTA DISTRICT

The Oneota district, also known as the Buena Vista district, is in Queen Canyon at the north end of the White Mountain Range 5 miles southeast of Queens, a station on the narrow-gage railroad that operates between Mina, Nev., and Keeler, Calif. Although part of the district is in Mineral County, the major part is in Esmeralda County, Nev.

The district was organized as early as 1862 by prospectors in search of gold in the White Mountains, but it was abandoned shortly after. In May 1870, some very rich float was found by an Indian, who showed it to William Wetherell. Wetherell returned with the Indian to the place where the float was discovered and located the Indian Queen mine, which became a small but prominent early-day producer. The ore was very rich, and in the early days some of it was hauled to Sacramento, Calif., by ox teams, a distance of several hundred miles. In 1882 a small stamp mill and White furnace were erected at the mouth of Queen Canyon to treat the ores from the Indian Queen mine. The ores were very base and required roasting.

The Poorman mine was discovered in this area in 1888, and in 1892 an apex suit was started between the Indian Queen and the Poorman Cos., which dragged on for 4 years. The animosity engendered by this suit reached such a point that actual battles took place. In 1896 the Indian Queen Co. won the suit and continued to operate until 1904. The district was idle until 1935, when the Ora Tahoma Mining Co., consisting of a group from the state of Washington, consolidated the principal holdings in the area, comprising 30 claims, two of which are patented. In 1936, the Ora Tahoma Mining Co. was driving several exploratory crosscuts in virgin ground. Six men were employed.

Complete statistics of production from the Oneota district are not available. Total production of the property has been in the neighborhood of $1,000,000. According to Whitehill[12]

> The Indian Queen Mine for the last three quarters of 1875 produced 896 1/2 tons of ore having a value of $132,682.96, an average of $148 per ton; and for the first half of 1876, 231 tons valued at $36,346.32, an average of $157.35 per ton.

The old mines were worked through tunnels. The Poorman tunnel is reported to have a length of 227 feet. The old workings were inaccessible at the time of writer's visit because they had caved.

The following extracts regarding the Indian Queen mine from early reports of the State Mineralogist are of interest.

> The vein matter is covered with a few feet only of loose earth and has a slate formation underneath. It has been opened up sufficiently to show a body of ore 400 feet in length, by 200 feet in width. It varies in thickness from a streak of colored ore to 4 feet. The ore contains a large proportion of galena, copper, and iron, with a compound of sulphide of silver and antimony. 25 tons of selected ore from this mine yielded $530 per ton.[13]
>
> The loose body of ore found here upon the surface, has been traced up to and into a well-defined vein, which has been opened to a depth of 60 feet, showing a vein 2 feet in thickness, which works, without sorting, from $200 to $400 per ton. The ore is very base, containing silver in almost every combination known to science, and can only be worked profitably by persons skilled and experienced in working base ores. The vein is found between granite and slate formations, and pitches to the east at an angle of 40°. The course of the vein is north and south.
>
> Three tons of ore are worked daily, which yield a profit of $1,000. For a small mine, this is undoubtedly the most productive in the State.[14]

Several mine dumps reported to contain a total of 150,000 tons were sampled by the Ora Tahoma Mining Co. and are said to average $3.26 per ton in gold and silver at current metal prices. A picked 10-pound sample of dump material taken by the writer assayed 29.2 ounces of silver and 0.07 ounce of gold. The sample showed quartz, galena, pyrite, and sphalerite. The tailings from the old mill at the mouth of the canyon have been scattered by cloudbursts.

PILOT MOUNTAINS DISTRICT

The Pilot Mountains or Sodaville district includes the southern part of the Pilot Mountains in southeast Mineral County. Mina, a town on the Southern Pacific R.R., is the supply center and nearest shipping point.

In 1913 cinnabar was discovered in the vicinity of Cinnabar Mountain by Thomas Pepper and Charles Keough while they were searching for stray steers. This discovery caused considerable excitement, and shortly afterward many claims were staked by men from Mina and Tonopah. The quicksilver deposits are in an area about 7 miles square on the northwest and southwest slopes of Cinnabar Mountain.

Quicksilver produced from various properties has amounted to more than $200,000, mostly from high-grade ore produced during the period from 1915 to 1917. Most of the production has been derived from properties of the Mina Mercury Co. and the Drew mine.

Cinnabar occurs in limestone, conglomerate, and silicified tuff. The geology of the deposits has been described by Foshag.[15]

In 1916 tungsten deposits were discovered on the east slope of Pilot Mountains 23 miles by automobile road from Mina, Nev. For several years after the discovery the mine was developed in a desultory manner, but a very small amount of tungsten concentrates were produced. The tungsten-bearing mineral is scheelite.

In addition to tungsten and quicksilver deposits, there are several deposits of gold and silver that have been intermittently productive for a number of years. In recent years metal mining in the area has been confined to small-scale operations.

A brief description of the geology of the tungsten deposits is given by Hess and Larsen.[16]

Mina Mercury Co.

The Mina Mercury Co., W. W. Booth, president, controls seven claims and three fractions of unpatented ground in Cinnabar Canyon. This property includes a part of the Lost Steers group, the original discovery claims located by Pepper and Keough. Property was last worked in 1929-30 under lease and option by the Nevada Almaden Quicksilver Co.

In 1929, this company replaced the four D-retorts with a Cottrell oilfired rotary furnace 30 feet long and 3 feet in diameter. Accessory reduction equipment includes a crusher, cyclone dust collector, metal-pipe condensing system, and redwood settling tanks. Power is furnished by 25-horsepower Fairbanks-Morse semi-Diesel engine.

Development includes a shallow inclined shaft and, to judge from the dump, a considerable amount of underground workings.

According to Foshag [17], the country rock is limestone, with which are interbedded thin layers of conglomerate, sandstone, and shale. The ore deposits are lenticular bodies in limestone. The richest ore is confined to a brecciated zone from 4 to 6 feet wide, which has a definite hanging wall but sometimes obscure footwall. Cinnabar is almost the only sulphide present, although some small patches of stibnite and a few grains of pyrite are found. The cinnabar forms seams and branches in the soft gangue or is finely disseminated through it. Beyond the sheared zone the cinnabar occurs as small scattered grains or bunches in seams of white calcite that cut the dark limestone.

Drew Mine

The Drew mine is at the head of Cinnabar Canyon about 14 miles easterly from Mina. Six unpatented claims and a millsite are owned by Mrs. Al Drew of Reno.

Development includes an incline shaft, said to be 300 feet deep, and considerable lateral workings. Most of the workings are caved and inaccessible. Most of the equipment has been removed.

According to Foshag[18], the cinnabar is present as irregular bodies in sandstone. The stopes mined are said to have been small, but the ore was rich; some of it carried as high as 18 percent quicksilver, 28 percent lead, 44 ounces silver, and more than 1 ounce gold to the ton. The ore is more complex here than in any other part of the district; the cinnabar is intimately associated with calamine with bintheimite. Fine-grained sphalerite is also abundant. The gangue is chiefly quartz.

Other Cinnabar Claims

Bert Hitt, of Mina, and partner own two cinnabar claims in Dunlap Canyon 11 miles from Mina. Very little development work has been done on this property and no quicksilver has ever been produced.

Adjoining the Hitt claims are two claims owned by Ed Allen of Mina. Property is developed by a 50-foot shaft and 350 feet of drifts and other workings. Production from the Allen property has been $7,000 in quicksilver.

The Mammoth Quicksilver Co. property in Dunlap Canyon comprises seven unpatented claims. Nobel C. Smith, of Fresno, Calif., is the president and principal owner of the company. Production is said to have been about 300 flasks.

Development consists of several tunnels and subsidiary workings totaling several thousand feet.

Equipment includes one D-retort and a three-pipe furnace, a small air compressor, an incline gravity tram 600 feet in length, and several camp buildings.

The Cardinal group of four claims, owned by Henry Ott of Reno, is near the Drew property in Cinnabar Canyon. This property was discovered in 1931 by J. R. Towner of Sodaville. Production is said to have been 70 flasks. The mine is equipped with two D-retorts.

The Red Wing group of 8 claims, owned by George A. Betty and Edward Messinger of Mina, is in the vicinity of Summit Springs, 23 miles from Mina.

Development work consists of several tunnels, which, with other workings, total 1,000 feet. Equipment includes two D-retorts. A small production of quicksilver has been made.

The Reward group of 10 claims, owned by George Thompson of Mina and associates, is in Dunlap Canyon. Production of quicksilver by lessees is said to have been about 1,000 flasks.

Development work comprises 300 feet of drifts, raises, and crosscuts. With the exception of about 40 feet of tunnel, all the work has been in ore. In October 1936, this property was under lease to Dave Hutchinson and Edward Messinger on a royalty basis of 15 percent of the production.

In October the lessees had constructed a two-pipe furnace with a capacity of about 3 tons of ore per day. This furnace consisted of two 12-inch pipes, 12 feet long, placed vertically in a brick-firing chamber. The furnace was fired with fuel oil.

The cinnabar is present in seams and disseminated in conglomerate, limestone, and sandstone.

The Chong Wong property is 1 1/2 miles northeast of the Drew mine. It comprises three claims owned by Chong Wong of San Francisco, Calif. Here the cinnabar is associated with barite in a chert formation.

In addition to the foregoing properties, other cinnabar claims are held in this area. The possibilities for further production of quicksilver from this area are promising.

Gunmetal Group

The Gunmetal group of claims, also known as the Summerfield property, is on the east slope of the Pilot Mountains 23 miles southeast of Mina. The principal owner is E. W. Esson of Los

Angeles. About 1924, this property was worked under bond and lease by the Lezeart Coal Mine Syndicate. This company erected a 25-ton mill that employed pneumatic concentration. In 1927, operations were discontinued. The attempt to recover the scheelite by dry concentration was unsuccessful. The Lezeart mill was equipped with a small crusher, an Abbe ball mill, and two Stebbins dry concentrating tables. Development consists of a tunnel about 600 feet long and some scattered surface workings.

The scheelite occurs in a series of metamorphosed limestone (tactite) beds that have been uplifted and tilted by granodiorite porphyry. The beds dip from 15° to 30° and range in width from 3 feet to a maximum of 75 feet. The tungsten mineral is scheelite which occurs as crystals that range in size from a head of a pin to 1/2 inch in diameter. The gangue minerals are garnet, epidote, and other silicates common to this type of contact metamorphic deposit. Sampling done has been insufficient to determine how much of the tactite beds can be classed as ore; by panning it is said most of the tactite is found to carry scheelite, and some of it will average 1 percent scheelite.

Other Tungsten Claims

The Desert Scheelite group of five unpatented claims, owned by George Thompson and M. Whitaker of Mina, adjoins the Summerfield property. In 1936 this group of claims was reported to have been taken over by a San Francisco syndicate on a royalty basis. Development work consists of short tunnels and surface cuts totaling about 500 feet.

The Garnet group of six unpatented claims in the same area is owned by George Thompson and the S. M. Summerfield estate. Virtually no development work has been done on this group.

The Silver Tungsten King group, comprising four claims, is owned by George Zark, of Mina, and associates. Very little work has been done on this group.

Stormland Group

The Stormland group of claims is at Camp Eddyville, in the Pilot Mountains, about 15 miles east of Sodaville. In 1932, John Eddy and Delbert Spainhour, of Mina, discovered high-grade gold ore on this ground. The property was sold to a group that is reported to have taken out $30,000 in high-grade gold ore by chloriding. In 1936 the property was inactive.

The formation is said to be limestone cut by dikes. Free gold is associated with manganese in irregular pockets in the limestone. All the ore produced has been taken from within 50 feet of the surface.

Belleville Mine

The Belleville mine comprises four unpatented claims at the head of Telephone Canyon on the west side of the Pilot Mountains 8 miles east of Sodaville. The owner is A. J. Belleville of Mina. The property was discovered in 1928 by Charles Woodruff. In 1932 and 1933 a small company called the Russell Gold Mining Co. operated the property and produced 827,000 in gold. In 1936, the property was under bond and lease to F. C. Marquiss and associates, and preparations were being made to erect a 25-ton mill.

Development consists of five tunnels, the longest of which is 900 feet. Total development is approximately 3,000 feet. Equipment includes a two-drill portable compressor and a 5-ton Straub mill for amalgamation. Two veins have been discovered on the Belleville property. Values are chiefly in free gold occurring in a gangue of manganese and calcite.

Sodaville Tailings

About 1/2 mile east of Sodaville is a tailings dump owned by Arthur Nelson of Candelaria, which was derived from ore taken from the mines of the Candelaria district in the early days. The ore was treated by chloridized roasting and amalgamation. A number of unsuccessful attempts have been made to recover the values. The dump is reported to contain at least 20,000 tons and to average better than $3.50 per ton at current metal prices. Values are in silver, gold, and quicksilver. The quicksilver in the tailings was derived from losses incurred in amalgamating the ores.

Bentonite

Two miles east of Mina on the west slope of the Pilot Mountains is a bentonite type of clay. The deposit is covered by two unpatented claims owned by the John McMillan estate. It has been prospected by several shallow shafts and open-cuts. No large tonnage is indicated.

Montezuma Mine

The Montezuma mine is in the east foothills of the Pilot Mountains, 20 miles by road east of Sodaville. Some years ago this property was worked by the German American Turquoise Co. In recent years the property has been in-active. Development consists of a number of irregular pits and short tunnels.

The turquoise is in decomposed trachyte as veinlets and nodules up to 1 inch thick. Most of the material is said to be of too poor quality for gem stones.

RAND DISTRICT

The Rand or Bovard district is on the northeast slope of the Gabbs Valley Range in northeast Mineral County, 27 miles by road northeast of Hawthorne via Ryan Canyon and Nugget Wash. It was discovered by Al Bovard and other prospectors from Rawhide in the spring of 1908. The nearest shipping point is Nolan on the Mina Hazen branch of the Southern Pacific R.R. 17 miles southwest of the district.

In 1919, a company called the Gold Pen Mines Co. acquired the Gold Pen property and erected a 20-ton amalgamation mill in 1920. This company ran into financial difficulties in 1921 and the property was sold to other interests.

Practically all the production has been ore shipped by lessees, except for small amounts treated at Rawhide or locally. Metal production from the district from 1910 to 1934 is shown in table 5.

Randall Property

R. J. Randall owns eight unpatented claims formerly owned by the Nevada-Rand Mining Co. Randall acquired the property in 1927 at a sheriff's sale to satisfy a judgment for $2,500 and costs obtained by the company's creditors. Property has been worked intermittently by Randall since 1927. Production is reported to have been about $50,000.

Development comprises a 1 1/2-compartment vertical shaft 450 feet deep and approximately 1,500 feet of underground workings. Equipment includes a gasoline hoist, blacksmith shop, and several camp buildings. Mining is done by hand methods. A shipment of ore made by Randall to the International Smelting & Refining Co. at Salt Lake City on April 12, 1935, afforded the following data:

I. C. 6941

TABLE 5. – Gold, silver, copper, and lead production from Rand (Bovard) District, Nevada, 1910-35, in terms of recovered metals

(Compiled by Charles White Merrill, Mineral Production and Economics Division, U. S. Bureau of Mines)

Year	No. of mines	Ore, short tons	Gold		Lode Silver		Copper		Lead		Total value	Average recoverable value of ore per ton/1
			Fine oz.	Value	Fine oz.	Value	Fine oz.	Value	Fine oz.	Value		
1910	4	12	74.85	$1,547	317	$171	---	---	---	---	$1,718	$143.17
1911	---	---	---	---	---	---	---	---	---	---	---	---
1912	---	---	---	---	---	---	---	---	---	---	---	---
1913	---	---	---	---	---	---	---	---	---	---	---	---
1914	3	227	398.76	8,243	18,230	10,081	---	---	---	---	18,324	80.72
1915	7	212	1,271.25	26,279	8,674	4,398	---	---	---	---	30,987	146.17
1916	9	812	566.85	11,718	4,746	3,123	---	---	6,588	$310	40,018	49.28
1917	9	2,795	105.90	2,189	2,390	1,969	100,709	$24,774	5,845	403	107,145	38.33
1918	6	265	70.87	1,465	2,927	2,927	374,395	102,210	9,034	777	17,615	66.47
1919	4	189	353.68	7,311	3,788	4,243	47,572	11,750	20,744	1,473	13,828	73.16
1920	3	95	743.84	15,376	1,793	1,954	12,223	2,274	---	---	17,330	182.42
1921	3	34	145.55	3,009	4,755	4,755	---	---	---	---	7,764	228.35
1922	4	41	121.02	2,502	1,191	1,191	---	---	---	---	3,693	90.07
1923	2	81	239.06	4,942	9,036	7,410	---	---	---	---	12,352	152.49
1924	4	370	759.28	15,696	33,678	22,564	---	---	---	---	38,260	103.41
1925	2	234	407.17	8,417	3,692	2,562	---	---	---	---	10,979	46.92
1926	4	477	250.18	5,172	4,650	2,902	26,141	3,660	---	---	11,734	24.60
1927	5	455	355.01	7,339	5,499	3,118	11,195	1,466	10,512	663	12,586	27.66
1928	6	243	182.12	3,765	2,465	1,442	11,242	1,619	---	---	6,826	28.09
1929	3	183	48.98	1,013	166	88	19,126	3,366	719	45	4,512	24.66
1930	---	---	---	---	---	---	---	---	---	---	---	---
1931	---	---	---	---	---	---	---	---	---	---	---	---
1932	---	---	---	---	---	---	---	---	---	---	---	---
1933	1	33	32.00	818	1,096	384	---	---	---	---	1,202	36.42
1934	1	22	27.30	954	813	526	---	---	---	---	1,480	67.27
1935	5	111	103.32	3,616	1,612	1,159	---	---	---	---	4,775	43.02
Total		6,891	6,256.99	$131,371	111,518	$76,967	602,603	$151,119	53,442	$3,671	$363,128	$52.70

1/ Not to be confused with average assay value of ore.

Metal quotation:
- Ag $0.64125 per oz.
- Au 35.00 per oz.

Settlement assay:
- Au 0.75 oz.
- Ag 22.25 oz.
- Zn .5 percent
- Cu nil
- Pb nil
- Inso 86.4 percent
- Fe 2.5 percent
- S .2 percent
- CaO percent

Metal payment:
Au 91 percent at $35 per oz.		$23.888
Ag 95 percent at $0.64125 per oz.		13.554
Gross value		37.442

Treatment charge:
Base rate	$4.00 per.ton		
10% charge metal payment over $25	1.244		
	5.244		5.244
Net value per ton			32.198

	Pounds		
Wet weight	96,780		
Less 1.54 moisture	1,490		
Dry weight	95,290	equals 47.645 tons	

47.645 tons at $32.198 per ton — $1,534.07

Deductions:
Freight	$4.90 per ton	$237.11	
Demurrage	2.00		
	239.11		239.11
Net proceeds			1,294.96

Ore occurs in veins in latite. Some of the silver is alloyed with gold and some occurs as cerargyrite and argentite. The rich ore occurs in lenses in a quartz gangue stained with iron and manganese oxides.

The ratio of gold to silver, by weight, averages 1 to 30.

Gold Pen Mine

The Gold Pen mine comprises a large group of claims owned by Richard Cowles of Reno. The property is developed by a shaft 250 feet deep, a tunnel, and several thousand feet of workings. Equipment includes an assay office, 25-ton mill, blacksmith shop, and camp building. In 1936, a portion of the property was under lease to W. B. Worlock, who had made two small shipments up to October.

At the Gold Pen mine the vein is said to be 3 to 8 feet wide and separated from the walls by sheets of alunite from 3 inches to 1 foot wide. Values are gold and silver in a quartz gangue. The enriched zone extends to a depth of 250 feet from the surface.

Lone Star Group

The Lone Star group of eight unpatented claims is owned by E. M. Mims, of Sacramento, Calif., and adjoins the Randall property. Development consists of a shaft 550 feet deep and about

1,000 feet of underground workings. Mining is done by hand.

Geological conditions on this property are essentially the same as on the Randall group. Production has been about $25,000 in gold and silver.

RAWHIDE DISTRICT

The Rawhide or Regent district is on an irregular range of hills at the south end of the Sand Springs Range in northeast Mineral County near the border of Churchill County. Fallon in Churchill County is 50 miles northwest, and Schurz, the nearest shipping point, 29 miles east.

Mining locations were first made in the area in 1906. With the finding of some high-grade gold ore and after considerable flamboyant advertising, a rush to the district took place in 1908 that raised the temporary population to 4,000 people. On September 4, 1908, a fire destroyed a large part of the business section of Rawhide, which caused a loss of several hundred thousand dollars. No large mines have ever been developed in this area and the principal production of the camp has been derived from leasing operations. In the early years of the camp as many as 50 sets of lessees or leasing companies were at work at one time. Nearly all the ore was shipped for treatment.

In 1909 two mills of unusual type were erected in the district. One was called a Tadmor and the other a Cannonball mill. The Tadmor mill employed a heavy muller actuated by an overhead eccentric bearing, which gave a crushing movement similar to that of a gyratory crusher. The muller weighed 8,000 pounds, and crushing was effected by the action of its weight upon a die. This mill is reported to have had a capacity of 20 tons per 24 hours. Concentrating was done with Wilfley tables.

The second mill is reported to have had a capacity of 30 tons per 24 hours, from 1 inch to 80-mesh, crushing either wet or dry. The crushing device consisted of a flat iron pan with three rings at increasing distances from the axis of rotation. Each ring carried a number of chilled iron balls, uniform in size for each ring but diminishing in size from the inner to the outer ring. The balls were held in place under pressure by overhead rings so designed that the pressure on each ring could be changed independently of the other two. The ore was fed to the center of the machine and was discharged at the periphery. Concentrating was done by amalgamation.

Both these mills operated only a short time and are mentioned here more because of their metallurgical interest than because of their economic importance. Several other mills using stamps for crushing were erected in the district in the early days. All of the mills have been dismantled.

The largest holdings in the district are owned by the Scheeline estate, which controls 17 1/2 patented claims, and the Grutt brothers, who own 19 claims, most of which are patented.

Development work consists of a number of shafts and tunnels, the deepest shaft being 600 feet with a winze from the bottom level 120 feet deep. Judging from the mine dumps, total underground workings will comprise at least several miles.

The formations in the area are rhyolite, dacite, and andesite. The ore occurs in a network of veinlets that impregnate the country rock. The mineralization appears to follow incipient fissures, mainly in kaolinized rhyolite. In places the kaolinite material carries good values. The ore minerals are native gold alloyed with silver, argentite, and cerargyrite.

At the time of the writer's visit in June 1936, several lessees were active in the camp.

Production of the district from 1908 to 1934 is shown in table 6.

Leonard Lease

For the past 18 years W. H. Leonard of Rawhide has shipped 15 to 20 tons of ore annually,

In this general view of rawhide looking east, the famous Stingaree Gulch bisects Grutt and Balloon Hill. Looming behind the crowds oon Rawhide's busy streets are several false fronted buildings housing mostly saloons.

TABLE 6 - Gold, silver, copper, and lead production from Regent (Rawhide) district, Mineral County, Nev., 1908-35

(Compiled by Charles White Merrill, Mineral Production and Economics Division, U. S. Bureau of Mines)

Year	No. of mines	Placer Gold Fine oz.	Placer Gold Value	Placer Silver Fine oz.	Placer Silver Value	Total value	No. of mines	Ore, short tons	Lode Gold Fine oz.	Lode Gold Value
1908	1	4.40	$91	---	---	$91	9	1,826	6,069.86	$125,475
1909	---	---	---	---	---	---	10	6,197	7,546.59	156,002
1910	5	579.79	11,985	467	$252	12,237	22	12,582	6,538.27	135,158
1911	7	324.25	6,699	271	143	6,842	17	6,013	3,364.65	69,554
1912	2	81.90	1,693	82	50	1,743	19	1,579	1,563.50	32,320
1913	1	8.20	170	10	6	176	13	9,954	4,845.44	100,164
1914	5	201.78	4,171	148	82	4,253	24	18,279	7,523.45	155,524
1915	3	274.70	5,679	166	84	5,763	5	6,973	3,968.80	82,042
1916	1	12.62	261	7	5	266	7	4,371	2,268.56	46,895
1917	2	17.22	356	12	10	366	5	298	423.19	8,748
1918	4	47.55	983	48	48	1,031	8	285	620.14	12,820
1919	5	45.34	937	28	31	968	5	199	677.97	14,015
1920	---	---	---	---	---	---	7	284	807.65	16,696
1921	2	6.60	136	5	5	141	11	93	257.10	5,315
1922	---	---	---	---	---	---	11	237	368.02	7,607
1923	1	2.04	42	1	1	43	8	399	393.62	8,137
1924	1	8.38	173	2	1	174	9	221	178.09	3,681
1925	---	---	---	---	---	---	6	36	123.02	2,543
1926	---	---	---	---	---	---	12	120	93.92	1,941
1927	---	---	---	---	---	---	5	57	119.09	2,462
1928	---	---	---	---	---	---	4	186	272.67	5,636
1929	---	---	---	---	---	---	3	66	82.09	1,697
1930	---	---	---	---	---	---	5	40	52.90	1,094
1931	2	18.29	378	10	3	381	6	116	103.87	2,147
1932	2	32.30	668	24	6	674	9	118	288.01	5,954
1933	5	65.93	1,685	29	10	1,695	8	279	122.94	3,142
1934	2	31.77	1,110	89	57	1,167	8	139	111.88	3,910
1935	4	54.64	1,912	42	30	1,942	8	100	248.26	8,689
Totals		1,817.70	$39,129	1,441	$824	$39,953		71,047	49,033.55	$1,019,368

Continued ———

TABLE 6. - _Gold, silver, copper, and lead production from Regent (Rawhide) district, Mineral County, Nev., 1908-35_ (Continued)

(Compiled by Charles White Merrill, Mineral Production and Economics Division, U. S. Bureau of Mines)

Lode (Continued)

Year	Silver		Copper		Lead		Total value	Average recoverable value of ore per ton[1]	Total value, lode and placer
	Fine oz.	Value	Pounds	Value	Pounds	Value			
1908	40,970	$21,714	---	---	---	---	$147,189	$80.61	$147,280
1909	89,833	46,713	---	---	---	---	202,715	32.71	202,715
1910	78,117	42,183	352	$45	216	$9	177,395	14.10	189,632
1911	44,740	23,712	---	---	---	---	93,266	15.51	100,108
1912	45,139	27,760	16,663	2,749	418	19	62,848	39.80	64,591
1913	109,817	66,329	38	6	---	---	166,499	16.73	166,675
1914	105,607	58,401	---	---	---	---	213,925	11.70	218,178
1915	78,512	39,806	---	---	---	---	121,848	17.47	127,611
1916	35,313	23,236	---	---	838	58	70,189	16.06	70,455
1917	10,397	8,567	---	---	---	---	17,315	58.10	17,681
1918	9,041	9,041	7,842	1,937	---	---	23,798	83.50	24,829
1919	12,972	14,529	---	---	---	---	28,544	143.44	29,512
1920	25,443	27,733	---	---	---	---	44,429	156.44	44,429
1921	9,528	9,528	---	---	---	---	14,843	159.60	14,984
1922	13,451	13,451	---	---	---	---	21,058	88.85	21,058
1923	9,924	8,138	---	---	---	---	16,275	40.79	16,318
1924	4,518	3,027	43	6	38	4	6,714	30.38	6,888
1925	4,324	3,001	20	3	19,470	1,558	5,551	154.19	5,551
1926	2,648	1,652	---	---	1,546	97	5,151	42.93	5,151
1927	686	389	689	90	---	---	3,038	53.30	3,038
1928	1,356	793	---	---	---	---	6,429	34.56	6,429
1929	1,168	623	107	14	530	26	2,320	35.15	2,320
1930	208	80	---	---	---	---	1,214	30.35	1,214
1931	507	147	---	---	---	---	2,294	19.78	2,675
1932	882	249	---	---	---	---	6,203	52.56	6,877
1933	472	165	---	---	310	12	3,307	11.85	5,002
1934	366	237	---	---	---	---	4,159	29.92	5,326
1935	1,425	1,024	37	3	---	---	9,716	97.16	11,658
Totals	737,364	$452,228	25,791	$4,853	23,366	$1,783	$1,478,232	$20.81	$1,518,185

[1] Not to be confused with average assay value of ore.

which had a value of from $50 to $1,200 per ton. In 1936 Leonard leased on the Truett property, owned by the Scheeline estate.

The royalty paid by Leonard on the net smelter returns is as follows:

Value of ore	Royalty, percent
$50 or less	10
$50 to $400	15
$400 or more	20

Mining is confined to working a large mass of low-grade ore that averages about $2.50 per ton in gold. According to Leonard, the gold occurs in a number of small fissures in dacite. On the surface the fissured area is 225 feet wide and 700 feet long. By screening the run of mine product through 1-inch screen, Leonard can obtain a product that averages about $100 per ton. Several m.n are employed, and mining is done by hand. Hoisting is done with a 25-horsepower Western gasoline hoist.

The smelter returns on a shipment of ore made by W. H. Leonard to the American Smelting & Refining Co. on October 18, 1935, gave the following returns:

Metal quotation:
    Au $34.9125
    Ag   .77

Settlement assay:
    Au   2.425 oz.
    Ag   7.7 oz.
    Insol.   88 percent

Metal payment:
    Au at $31.81825 $77.16
    Ag 95% less 0.5 oz. at .77   5.27

Treatment Charge
    base   3.75
    10% excess over $20.00   2.25 (max)
    6.00   6.00
    Net value per ton   76.43

	Pounds		
Wet weight	45,040		
Moisture, less 4.2%	1,892		
Dry weight	43,148	21.574 tons at $76.43	$1,648.90
Royalty before hauling, 15%		$214.57	
Freight at $9.30		209.44	
Emergency freight at $0.40 per ton		9.01	
Hauling at $3.50 per ton		78.82	511.84
Net proceeds			1,137.06

Placers

Placer gold has been found south and southwest of the town of Rawhide. The lack of water necessitated the use of dry-washing equipment, although hand rockers were employed to some extent in the early days. Water for recovering gold with rockers was hauled from wells several miles southeast of the camp.

The best placer diggings were on the southeast slope of Hooligan Hill. Numerous shafts averaging about 15 feet in depth attest the activity in his area in former days. The gravel is composed of angular rock fragments, sand, and soil, with few large boulders. The largest nugget found is reported to have had a value of $70.

Hooligan Hill slopes toward a canyon several miles in length, and at the mouth of this canyon the gravels spread out in the form of an alluvial fan, in which placer gold has been found. The gravels range in depth from 40 to 90 feet. About 30 shafts have been sunk on the fan, and a small amount of placer mining has been done at several of the shafts with small-scale placer mining equipment. The best values are found directly above the rhyolite bed-rock. In 1930 the Idaho Dredging Co. of Boise, Idaho, obtained a bond and lease on 1,180 acres of placer ground and began sampling. After a short time the property and the sampling operations were taken over by the Hammond Engineering Co. This company relinquished the option on the ground. In 1935 Rene Engel and associates, of Pasadena, Calif., relocated 15 claims, called the Pilot placers, in this area.

In recent years there has been a small amount of dry washing of placer ground in the vicinity of Rawhide by itinerant placer miners. The returns of this work netted the operators less than wages.

Tungsten Claims

In 1930 W. H. Leonard discovered scheelite 5 1/2 miles east of Rawhide in the southern part of the Sand Springs Range. In 1936 the Leonard group of 10 unpatented claims was under bond and lease to the Mills Alloys, Ltd., of Los Angeles, Calif. In September 1936 this company erected a small concentrator at Dead Horse Wells 9 miles south of the deposit, where water is available. At the time of the writer's visit this company had produced 600 pounds of concentrates carrying about 40 percent scheelite. The concentrates had not been cleaned by magnetic separation and contained a considerable amount of garnet, This is the only production from the property.

Development work on the claims is all superficial in character and consists of approximately 300 feet of scattered open-cuts and short tunnels. The Mills Alloys, Ltd., employed four men, and mining was done by hand methods.

Scheelite occurs as a contact metamorphic deposit in limestone near granite. The tactite, which is the garnetized rock that carries the scheelite, shows a considerable width in places and is traceable for several thousand feet on the surface. The scheelite is in the form of small crystals from the size of a pinhead up to half an inch in diameter. In places the tactite is stained with copper. No systematic sampling has been done, and the property may be considered to be in the prospect stage.

The concentrator built by the Mills Alloys, Ltd., consists of a small jaw crusher, two sets of small rolls, one concentrating table, and a home-made electric magnetic separator for cleaning the table concentrates. Power is furnished by a gasoline engine. None of the equipment is of standard make, and the capacity of the plant is limited to a few tons of ore per day.

Several miles north of the Leonard property several tungsten claims are owned by Tom Kenyon of Sunnyside, Nev. This property was not visited, but specimens of the ore seen by the writer showed scheelite as crystals ranging in diameter from 1/4 up to 1 inch in a matrix of amphibole and quartz. It is reported that very little work has been done on the property.

RHODES MARSH DISTRICT

Rhodes Marsh is in the lowest part of Soda Springs Valley just east of Rhodes siding on the Tonopah & Goldfield R. R. The town of Mina is 8 miles north. The altitude of the surface of the marsh is 4,400 feet above sea level.

The marsh was first exploited in the early sixties to supply salt necessary for the extraction of gold and silver from the ore of the Comstock, Aurora, Candelaria, Belmont, and other districts where the Washoe pan process[19] was employed.

An interesting sidelight on the production of salt from Rhodes Marsh is a historical note, mentioning the importation of camels for transporting salt to Virginia City. An old ordinance in Virginia City prohibited camels from entering the city during certain hours of the day to prevent the stampeding of horses.

In the seventies, shortly after its discovery in Teel's Marsh, borax was found in Rhodes Marsh. Considerable activity prevailed and borax mining was pursued for a number of years. The borax occurred mainly as ulexite nodules consisting of rounded masses of loosely compacted acicular crystals several inches in diameter, called "cottonballs" because of their resemblance to balls of cotton. Because these deposits are rather purer than those of the natural borax type, they were much sought after and exploited. The ulexite nodules occurred a few inches below the surface of the lake and were gathered by Chinese and Indians, who hand-picked them out of the shallow excavations in the mud. These nodules were boiled in tanks fired with sage- brush and pinon pine. After the material was in solution, the borax was crystallized out on iron rods suspended in the crystallizing vats. The crude borax was shipped to Alameda, Calif., for refining.

Rhodes Marsh was thoroughly worked for borax in the early days, and it ceased to be a factor when the extensive underground deposits of calcium borate (colemanite) found in the vicinity of Death Valley, Calif., revolutionized the industry.

Some years ago the marsh was prospected for potash, but the potash content of the brines present a short distance beneath the surface is too low to be of commercial interest.

In 1928, P. S. Williams, a chemical engineer, became interested in Rhodes Marsh as a source of sodium sulphate. For many years it was known that mirabilite (Glauber's salt, $Na_2SO_4 \cdot 10 H_2O$) occurred in the marsh in large quantities, but nothing was done until Williams interested a group of San Francisco men in obtaining an option on 2,120 acres owned by George A. Betty of Mina, Nev. The San Francisco group organized a company called the Rhodes Alkali & Chemical Corporation and spent considerable money in sampling and investigating processes for the recovery of the sodium sulphate. In the first plant erected in 1930 efforts were directed toward the utilization of seasonal temperature changes in the production of sodium sulphate with natural brines made by passing water over or through the sodium sulphate beds. Next, an attempt was made to free the Glauber's salt from the silt with which it is associated, and then to dehydrate it by heating. Neither process.was successful. Later it was found that considerable thenardite (Na_2SC_4) occurred in the marsh and attention was directed to the mining of this product. In 1932, a plant capable of producing 150 tons of sodium sulphate per day was erected and was operated up to 1933.

The Rhodes Alkali & Chemical Corporation is said to have spent $150,000 in sampling and in the erection of a plant. Production of sodium sulphate has been about 20,000 tons.

Rhodes Marsh is 3 miles long and has an average width of 1 1/2 miles. It is dry for the greater part of the year and encrusted with salts. The accumulation of surface salts, mostly sodium chloride, varies in thickness with the season; during warm, dry weather the whole surface is elevated several inches and forms a pulverent mass; this mass shrinks again in cold, wet weather, becoming hard and compact. It rises and falls somewhat in the manner of a great pan of dough. The water level fluctuates with the season and in the summer stands 4 to 5 feet below the surface. The thenardite occurs in lenses from 3 to 5 feet thick under an overburden of silt and salt varying from 2 to 6 feet thick. The Rhodes Alkali & Chemical Corporation has sampled the marsh with several thousand augur holes drilled by hand, and, according to the drill records, at least 3,000,000 tons of sodium sulphate salts are available.

In mining, the thenardite was stripped of the overburden of salt and silt with a P. & H. gasoline shovel. The areas stripped are from 30 to 50 feet in diameter. These pits were drained by means of gasoline pumps.

The thenardite layer was blasted and loaded into 2-yard dump cars by gasoline shovel equipped with trench hoe attachment. The material was hauled about 1 mile to the plant for treatment. A flow sheet of the treatment plant is shown in figure 3.

The treatment of the crude thenardite is merely a process for the removal of sand silt. This is done in several steps to maintain a counter flow of material and wash liquors and to utilize the nearly saturated solu- tion so as to prevent losses. The following is a typical analysis of the finished product:

	Percent
Sodium sulphate	97.33
Sodium chloride	1.07
Calcium sulphate	.28
Insoluble	1.24
Moisture	.08
Total	100.00

Power for the treatment plant is furnished by a 120-horsepower Fairbanks-Morse Diesel engine.

Water is available from shallow wells near the plant. Two artesian wells were drilled by the company at the south end of the marsh in 1930. One well is 450 feet and the other 420 feet deep; both wells have 16-inch casing near the surface and 10-inch casing near the bottom. .n sinking the wells, three fresh-water-bearing strata were encountered. The combined flow from the two wells is in excess of 200 gallons per minute.

SANTA FE DISTRICT

The Santa Fe, also known as the Luning district, is in the northern part of the Pilot Mountains 4 miles east of Luning, a small town on the Mina-Hazen branch of the Southern Pacific R. R. The district is separated from the Gabbs Valley Range on the north by the Santa Fe Pass, which transects the range from east to west.

According to Burchard[20], the Santa Fe mine was discovered in 1879 and was in operation in 1883. The Sunrise mine also was active at that time; a little later the Wall Street mine was worked for copper. The ores from the Santa Fe and Sunrise mines are said to have carried rich silver values.

Up to 1894 mining in this area was confined mainly to the silver deposits; the copper-lead deposits, carrying some silver, have been worked intermittently, principally by lessees, from 1900 to 1929. The greatest period of activity occurred during the World War, when the high price of copper enabled lessees to work the deposits at a profit. Several small companies were organized to work properties in the district, but for the most part they were short-lived.

By selective mining and hand sorting, lessees were able to produce ore containing from 5 to 12 percent copper. The copper ore produced was shipped either to smelters in Salt Lake Valley, Utah, or to the Thompson smelter at Wabuska, Nev. The Thompson smelter was blown in during 1912 and closed down in 1928.

In 1936 mining activity in the Santa Fe district was confined to several gold deposits about 5 miles northeast of Luning.

The geology of the district has been described by Hill[21] and Clark.[22] Production of the district from 1906 to 1935 is shown in table 7.

New Year Group

The New Year group comprises four unpatented claims on the west slope of Pilot Range 5

TABLE 7. – Gold, silver, copper, and lead production from Santa Fe district, Mineral County, Nev., 1906-35
(Compiled by Charles White Merrill, Mineral Production and Economics Division, U. S. Bureau of Mines)

Year	Placer					Lode					
	No. of Mines	Gold		Silver		Total Value	No. of Mines	Ore, short tons	Gold		Silver	
		Fine oz.	Value	Fine oz.	Value				Fine oz.	Value	Fine oz.	Value
1906	---	---	---	---	---	---	4	7,000	1,685.24	$34,837	2,857	$1,943
1907	---	---	---	---	---	---	4	9,489	2,023.53	41,830	8,056	5,317
1908	---	---	---	---	---	---	8	2,143	713.10	14,741	3,102	1,644
1909	---	---	---	---	---	---	7	409	261.42	5,404	19,088	9,926
1910	---	---	---	---	---	---	8	1,120	151.08	3,123	10,022	5,412
1911	---	---	---	---	---	---	4	158	57.63	1,191	2,902	1,538
1912	---	---	---	---	---	---	34	3,080	352.14	7,279	17,415	10,710
1913	---	---	---	---	---	---	29	9,087	76.22	1,576	14,358	8,672
1914	1	11.56	$239	2	$1	$240	19	1,426	20.51	424	4,313	2,385
1915	---	---	---	---	---	---	20	2,726	144.47	2,987	6,802	3,449
1916	---	---	---	---	---	---	19	17,665	59.09	1,222	46,755	30,765
1917	---	---	---	---	---	---	49	19,932	97.17	2,009	50,007	41,206
1918	---	---	---	---	---	---	25	12,897	46.09	953	42,012	42,012
1919	---	---	---	---	---	---	4	534	.14	3	2,008	2,249
1920	---	---	---	---	---	---	5	157	8.50	176	1,717	1,871
1921	---	---	---	---	---	---	4	206	249.17	5,151	1,600	1,600
1922	---	---	---	---	---	---	4	204	159.96	3,307	2,079	2,079
1923	---	---	---	---	---	---	1	37	1.31	27	527	432
1924	---	---	---	---	---	---	2	13	1.22	25	319	214
1925	---	---	---	---	---	---	3	50	4.71	97	2,469	1,714
1926	---	---	---	---	---	---	---	---	---	---	---	---
1927	---	---	---	---	---	---	7	175	35.44	732	1,736	984
1928	---	---	---	---	---	---	5	350	188.58	3,898	1,809	1,058
1929	---	---	---	---	---	---	5	317	28.15	582	1,358	724
1930	---	---	---	---	---	---	1	7	18.37	380	23	9
1931	---	---	---	---	---	---	---	---	---	---	---	---
1932	---	---	---	---	---	---	---	---	---	---	---	---
1933	---	---	---	---	---	---	---	---	---	---	---	---
1934	---	---	---	---	---	---	2	145	49.53	1,731	518	335
1935	---	---	---	---	---	---	3	219	90.18	3,156	170	122
Totals	1	11.56	$239	2	$1	$240		89,546	6,522.95	$136,841	241,022	$178,370

Continued

TABLE 7. - Gold, silver, copper, and lead production from Santa Fe district, Mineral County, Nev., 1906-35 (Continued)

(Compiled by Charles White Merrill, Mineral Production and Economics Division, U. S. Bureau of Mines)

Year	Lode (Continued)						Average recoverable value of ore per ton [1]	Total value, lode and placer
	Copper		Lead			Total value		
	Pounds	Value	Pounds	Value				
1906	---	---	---	---		$36,780	$5.25	$36,780
1907	105,199	$21,040	64,000	$3,392		71,579	7.54	71,579
1908	24,334	3,212	19,237	808		20,405	9.52	20,405
1909	34,877	4,534	20,000	860		20,724	50.67	20,724
1910	3,521	447	23,288	1,025		10,007	8.93	10,007
1911	1,256	157	26,393	1,188		4,074	25.78	4,074
1912	311,559	51,407	13,749	619		70,015	22.73	70,015
1913	646,812	100,256	6,933	305		110,809	12.19	110,809
1914	190,193	25,296	1,240	48		28,153	19.74	28,393
1915	384,258	67,245	3,468	163		73,844	27.09	73,844
1916	2,547,053	626,576	---	---		658,563	37.28	658,563
1917	2,590,756	707,276	35,799	3,079		753,570	37.81	753,570
1918	1,923,259	475,045	18,011	1,279		519,289	40.26	519,289
1919	85,470	15,897	---	---		18,149	33.99	18,149
1920	1,048	187	19,966	1,597		3,831	24.40	3,831
1921	27	4	905	41		6,796	32.99	6,796
1922	372	50	3,948	217		5,653	27.71	5,653
1923	---	---	---	---		459	12.41	459
1924	48	6	5,971	478		723	55.62	723
1925	---	---	600	52		1,863	37.26	1,863
1926	---	---	---	---		---	---	---
1927	22,310	2,923	4,657	293		4,932	28.18	4,932
1928	6,251	900	---	---		5,856	16.73	5,856
1929	29,130	5,127	---	---		6,433	20.29	6,433
1930	---	---	---	---		389	55.57	389
1931	---	---	---	---		---	---	---
1932	---	---	---	---		---	---	---
1933	---	---	---	---		---	---	---
1934	---	---	---	---		2,066	14.25	2,066
1935	51	4	---	---		3,282	14.99	3,282
Totals	8,907,759	$2,107,589	268,165	$15,444		$2,438,244	$27.23	$2,438,484

[1] Not to be confused with average assay value of ore.

I. C. 6941

miles northeast of Luning, owned by H. A. Peterson and Joe Cardwell of Mina, Nev.

These claims formerly were part of the property owned by the Luning Consolidated Silver Mines Co. of Nevada. In 1936, lessees shipped several hundred tons of ore from this property to a Salt Lake City smelter.

Development work consists of a shaft 40 feet deep, another 175 feet deep, and several tunnels, comprising in all several thousand feet of under-ground workings. Mining is done by hand methods.

The gold occurs in a vein in limestone. The dip of the vein averages 50 degrees and the width about 5 feet. Commercial ore is present in lenses associated with pyrite in a quartz gangue. Near the surface the oxidation of the sulphide has resulted in the concentration of gold in a limonitic quartz gangue.

A shipment of ore made by E. S. Perry to the American Smelting & Refining Co. on April 3, 1936, afforded the following data:

Metal quotations:	Au	$34.9125 per oz.	
	Ag	.77 per oz.	
Settlement assay:	Au	.42 oz.	
	Ag	4.6 oz.	
	Cu	.18 percent	
Metal payment:	Au at $31.81825	$13.36	
	Ag less 0.5 oz at $0.77	3.16	
	Gross value	16.52	
	Less treatment	3.50	
		13.02	
	Pounds		
Wet weight	102,960		
Less 7.4% moisture	7,620		
Net weight	95,340 or 47.67 tons at $13.02		$620.66
Deductions:	Freight 3.60 per ton	$185.33	
	Emergency freight 7%	12.97	
	Hauling at $1.25 per ton	64.35	262.65
	Net proceeds		358.01

Dolly Group

The Dolly Group of four unpatented claims is owned by W. van Schoick, of Luning, Nev., and associates. This property is south of the New Year group and was formerly called the Luning Consolidated Gold Mining Co.

The property is developed by shafts and tunnels. The deepest shaft is 250 feet deep and the longest tunnel is 900 feet long. Underground workings total about 2,500 feet. Ore is mined by hand methods.

In 1935 the present owners erected a small amalgamation concentration mill at Luning, and up to June 1936 about 400 tons of ore, averaging $15 per ton, had been treated.

Mill equipment includes a jaw crusher 6 by 9 inches, a Sacramento mill [23], an amalgamation plate 4 1/2 feet wide and 5 feet long, and a concentrating table. By grinding to 50-mesh a recovery of 75 to 80 percent is obtained by amalgamation and table concentration.

Power for milling is furnished by a 50-horsepower Waukesha engine. Water for milling is pumped from a well near the millsite.

According to van Schoick, the formations on the Dolly group are granite and diorite. Gold ore is present in several fissure veins that range in width from 18 inches to 4 feet and dip 35 to 50°.

American Copper Co.

The most productive copper properties in the Santa Fe district were the Wall Street and Turk mines in New York Canyon 7 miles east of Luning. About 1929 these two mines and other properties in the locality were incorporated as the American Copper Co. Holdings of this company comprise 12 unpatented and 2 patented claims.

Development comprises several shafts, tunnels, and lateral workings that total about 2 miles. The deepest shaft is 428 feet. According to Hill[24], the country rocks consist mainly of crystalline limestones, probably of Triassic age, that have been intruded by granitoid rocks that range from quartz monzonite to quartz diorite and probably are Cretaceous.

At the Wall Street mine copper carbonate ore is present in an eastward striking brecciated zone 30 feet wide in westward-dipping limestones. The principal ore minerals are malachite, chrysocolla, azurite, and cuprite, which occur in thin irregular masses in reddish jasperoidal limestone.

Other Mines and Prospects

A number of other properties in the vicinity of New York Canyon produced small quantities of ore in former years, including the Champion, Mayflower, Mastodon, Silver Guardian, Vacation, Neversweat, Wedge, Copper Queen, Ideal, Giroux, and Calvada. These properties have been prospected and worked through tunnels and shafts ranging from 150 to 800 feet in depth.

The commercial copper ore is composed of copper carbonates and oxides in limestone, occurring in masses of irregular size and shape. All these properties have been idle for many years, and most of the mine equipment has been removed. The opportunities for mining additional copper ore from these properties are promising if the price of copper ever attains a figure comparable to the World War price. With the present price of copper the outlook for producing ore of shipping grade is not encouraging.

SILVER DISTRICT

The Silver Star, also known as the Gold Range, Mina, or Douglas district, is in Excelsior Mountains in southern Mineral County. The nearest shipping point is Mina on the Mina-Hazen branch of the Southern Pacific R. R., 6 miles northeast of Camp Douglas. The Douglas portion of the district covers a mineralized area roughly 2 miles long and 1 mile wide.

Veins carrying gold and silver were discovered at Camp Douglas by Pepper, Grassi, and Robb in 1893. From 1893 to 1903 considerable activity prevailed at Camp Douglas, and during this period lessees are reported to have produced about $500,000. During the panic of 1893, Camp Douglas was known as the "Dinner Pail", because of the opportunity afforded lessees to make a good living. The discovery of bonanza ore at Tonopah in 1900 drew many of the leasers away from the camp, so that production declined.

In former years, several small mills were erected in the Douglas area, but the bulk of the ore has been shipped to smelters for treatment. Most of the ore has been mined from shallow depths; the deepest working is the Bounce shaft, 425 feet deep.

The metal production of the Silver Star district from 1902 to 1935 is shown in table 8.

I. C. 6941

TABLE 8. — Gold, silver, copper and lead production from Silver Star district, Mineral County, Nev., 1902-35
(Compiled by Charles White Merrill, Mineral Production and Economics Division, U. S. Bureau of Mines)

Year	Placer						Lode			
	No. of mines	Gold		Silver		Total value	No. of mines	Ore, short tons	Gold	
		Fine oz.	Value	Fine oz.	Value				Fine oz.	Value
1902	---	---	---	---	---	---	1	480	288.41	$5,962
1903	---	---	---	---	---	---	3	205	359.33	7,428
1904	---	---	---	---	---	---	4	256	225.52	4,662
1905	---	---	---	---	---	---	5	697	718.37	14,850
1906	---	---	---	---	---	---	2	650	290.25	6,000
1907	---	---	---	---	---	---	---	---	---	---
1908	---	---	---	---	---	---	5	234	108.01	2,233
1909	---	---	---	---	---	---	4	127	41.75	863
1910	---	---	---	---	---	---	7	250	208.11	4,302
1911	---	---	---	---	---	---	7	121	65.42	1,352
1912	---	---	---	---	---	---	17	1,505	344.17	7,115
1913	---	---	---	---	---	---	14	413	258.64	5,347
1914	---	---	---	---	---	---	8	296	181.58	3,754
1915	---	---	---	---	---	---	21	2,016	1,528.49	31,597
1916	---	---	---	---	---	---	20	1,485	312.56	6,461
1917	---	---	---	---	---	---	22	2,651	107.27	2,218
1918	---	---	---	---	---	---	18	2,262	40.16	830
1919	---	---	---	---	---	---	6	439	8.64	179
1920	---	---	---	---	---	---	7	118	41.70	862
1921	---	---	---	---	---	---	5	50	18.39	380
1922	---	---	---	---	---	---	6	133	84.93	1,755
1923	---	---	---	---	---	---	8	143	96.95	2,004
1924	---	---	---	---	---	---	7	345	196.80	4,068
1925	---	---	---	---	---	---	9	390	472.50	9,767
1926	---	---	---	---	---	---	8	333	266.58	5,511
1927	1	21.34	$441	5	$3	$444	10	236	188.15	3,889
1928	---	---	---	---	---	---	8	538	323.53	6,688
1929	---	---	---	---	---	---	7	322	416.38	8,607
1930	---	---	---	---	---	---	3	583	309.74	6,403
1931	---	---	---	---	---	---	6	1,232	736.37	15,222
1932	---	---	---	---	---	---	12	886	549.97	11,369
1933	---	---	---	---	---	---	11	739	610.28	15,599
1934	---	---	---	---	---	---	17	5,083	1,804.71	63,075
1935	---	---	---	---	---	---	14	7,443	2,876.06	100,662
Total		21.34	$441	5	$3	$444		32,220	14,079.72	$361,014

Continued —

TABLE 8. – Gold, Silver, copper and lead production from Silver Star district, Mineral County, Nev., 1902-35 (Continued)
(Compiled by Charles White Merrill, Mineral Production and Economics Division, U. S. Bureau of Mines)

	Lode (Continued)								Total value, lode and placer
	Silver		Copper		Lead		Total value	Average recoverable value of ore per ton[1/]	
Year	Fine oz.	Value	Pounds	Value	Pounds	Value			
1902	148	$78	---	---	---	---	$6,040	$12.58	$6,040
1903	298	161	---	---	---	---	7,589	37.02	7,589
1904	19,008	11,025	---	---	---	---	15,687	61.28	15,687
1905	3,356	2,047	56,376	$8,795	14,106	$663	26,355	37.81	26,355
1906	5,124	3,484	180,000	34,740	120,000	6,840	51,064	78.56	51,064
1907	---	---	---	---	---	---	---	---	---
1908	1,345	713	591	78	64,453	2,707	5,731	24.49	5,731
1909	11,915	6,196	315	41	45,581	1,960	9,060	71.34	9,060
1910	7,660	4,142	536	68	15,427	679	9,191	36.76	9,191
1911	4,829	2,559	35	4	6,103	275	4,190	34.63	4,190
1912	7,663	4,713	100,244	16,540	30,478	1,371	29,739	19.76	29,739
1913	3,835	2,316	12,512	1,939	12,621	555	10,157	24.59	10,157
1914	2,365	1,308	947	126	38,634	1,507	6,695	22.62	6,695
1915	14,843	7,525	41,138	7,199	181,031	8,508	54,829	27.20	54,829
1916	8,252	5,430	190,567	46,880	7,906	545	59,316	39.94	59,316
1917	9,038	7,447	283,473	77,388	24,033	2,067	89,120	33.62	89,120
1918	4,415	4,415	225,417	55,678	9,812	697	61,620	27.24	61,620
1919	3,070	3,438	31,335	5,828	898	48	9,493	21.62	9,493
1920	11,687	12,739	1,079	199	71,378	5,710	19,510	165.34	19,510
1921	4,529	4,529	490	63	7,444	335	5,307	106.14	5,307
1922	8,690	8,690	93	13	19,363	1,065	11,523	86.64	11,523
1923	2,849	2,336	1,219	179	16,495	1,154	5,673	39.67	5,673
1924	503	337	31	4	5,537	443	4,852	14.06	4,852
1925	3,329	2,310	1,229	175	22,955	1,997	14,249	36.54	14,249
1926	10,651	6,646	6,502	910	30,885	2,471	15,538	46.66	15,538
1927	3,035	1,721	4,364	572	12,222	770	6,952	29.46	7,396
1928	1,053	616	66	10	6,611	384	7,698	14.31	7,698
1929	2,525	1,346	242	43	12,567	792	10,788	33.50	10,788
1930	561	216	---	---	---	---	6,619	11.35	6,619
1931	1,061	308	---	---	---	---	15,530	12.61	15,530
1932	2,337	659	248	16	3,489	105	12,149	13.71	12,149
1933	980	343	160	10	---	---	15,952	21.59	15,952
1934	3,552	2,296	97	8	6,854	254	65,633	12.91	65,633
1935	5,690	4,090	316	26	9,119	365	105,143	14.13	105,143
Total	170,205	$116,179	1,139,622	$257,532	796,002	$44,267	$778,992	$24.21	$779,436

1/ Not to be confused with average assay value of ore.

The tungsten deposits in the Excelsior Mountains were discovered in 1916 by Charles W. Noble on claims that he originally located for silver. Shortly after the discovery of tungsten the Noble property was sold to Atkins-Kroll Co., who operated it until 1918, when the mine was sold to other interests.

General Tungsten Corporation

The General Tungsten Corporation, incorporated in 1926, George F. Thompson, president, owns 10 unpatented claims in the Excelsior Mountains, 14 miles southwest of Mina via Pepper Springs. From Pepper Springs there is no road to the property, and supplies and ore must be transported by pack train.

Development consists of two tunnels, respectively 600 and 300 feet in length, a shaft 50 feet deep, and several open-cuts, totaling in all about a thousand feet. There is no equipment on the property.

Production has been one carload of gold-silver ore shipped in 1926 and 350 tons of scheelite ore milled at the Wagner mill at Silver Dyke in 1928. The scheelite ore is said to have carried 3 1/2 percent WO_3. In 1936, the corporation applied for a R. F. C. class B mining loan of $20,000 for further development of the property. It was stated that this loan has been approved.

The Silver Dyke vein system traverses the property and in places is 100 feet wide. Values are in gold, silver, and scheelite in a gangue of quartz and country rock.

Silver Dyke Mine

The Silver Dyke mine and concentrator are on the east flank of the Excelsior Mountains 14 miles by automobile road southwest of Mina, Nev. The Silver Dyke property comprises nine unpatented claims controlled by the Nevada-Massachusetts Co., Inc., which also owns tungsten mines near Mill City, Nev. During the World War period this property was owned and operated by the Atkins-Kroll Co. of San Francisco. This company erected a 50-ton concentrator at Sodaville for the treatment of the scheelite ore. In 1926, lessees constructed a 25-ton concentrator on the Wagner property, which was operated until 1927. In 1929 the Wagner and Silver Dyke properties were acquired by the Nevada-Massachusetts Co., Inc., and in 1930 the present concentrator of 60 tons capacity was erected.

The value of the production of scheelite concentrates from this property has been close to $1,000.000.

In 1936, the Silver Dyke mine and concentrator employed a crew of 35 men, and 55 tons of scheelite ore per day were being treated. The concentrates produced averaged between 63 and 66 percent WO_3.

Development includes a crosscut tunnel 800 feet long and several shafts totaling, in all, about 4,000 feet of underground workings. The plan and section of the mine workings are shown in figure 4. The ore is mined by shrinkage and open-stope mining methods.

Mining equipment includes an Ingersoll Rand Imperial-type 10-compressor, rock drills, steel sharpener, and a Baldwin Westinghouse storage-battery locomotive.

The scheelite ore is concentrated by tabling, following by magnetic separation to clean the bulk table concentrates. The flow sheets of the crushing plant and concentrator are shown in figures 5 and 6. Mining cost is $4.50 and milling $2 per ton.

Electric power for mining and milling is purchased from the Mineral County Power System. Power consumption averages 60,000 kilowatt-hours per month at a cost of about 2 cents per kilowatt-hour.

Water for milling is obtained from Spearmint Springs in Spring Canyon in the Pilot Range.

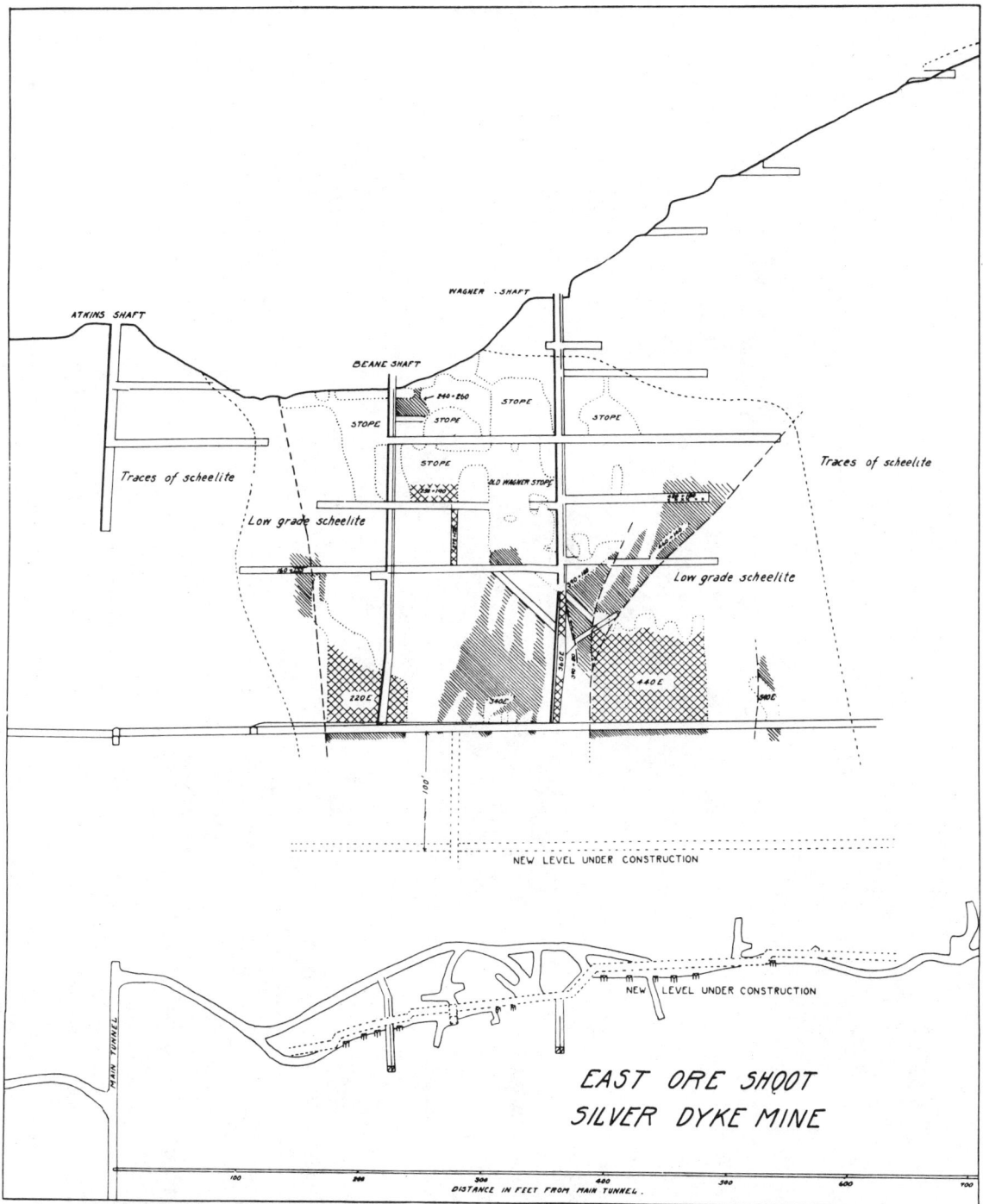

Figure 4.— Plan and section of Silver Dyke mine, Mineral County, Nev. (After Kerr.)

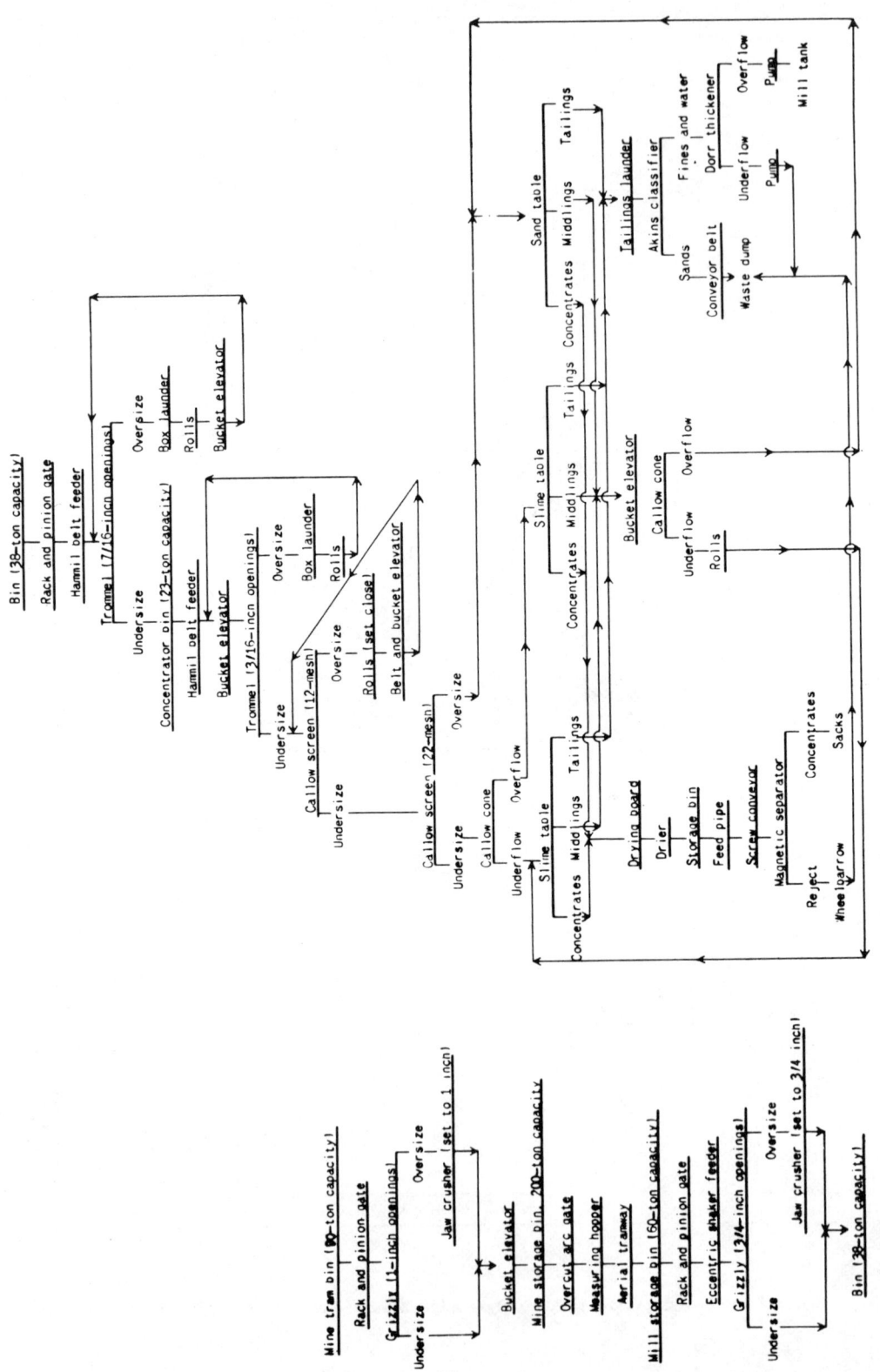

Figure 6.— Flow sheet of Silver Dyke concentrator.

Figure 5.— Flow sheet of Silver Dyke crushing plant.

The pipe line is 10 1/4 miles in length and 4 inches in diameter and is buried a depth of 30 inches. The flow of water to the mine is by gravity and pumping. Pumping lift is 1,528 feet, and the water is raised by three Ingersoll Rand, direct-connected, centrifugal pumps automatically controlled. In December 1932 a major earthquake in the vicinity of Mina cut off the mill water supply formerly obtained from a well and springs near the mine.

The geology of the Silver Dyke mine has been described by Kerr[25].

The Silver Dyke vein system has a length of 4 1/2 miles and a maximum width of 600 feet. This vein system has been called the Great Silver Dyke of western Nevada, although it is a vein rather than a dike and the silver mineralization has not been important from the production viewpoint.

In a general way the vein system follows a contact between diorite and other volcanics. The scheelite-bearing veins measure from a few inches to 15 feet in thickness. The ore shoots are spaced irregularly along the vein system and range from 60 to 120 feet in length. The scheelite, in grains and masses, occurs in quartz associated with albite. A small amount of pyrite is present also. The ore mined averages about 1 percent scheelite.

Tungsten Dike Group

The Tungsten Dike group of six unpatented claims, owned by E. T. Heggland of Mina, is in the Excelsior Mountains 12 miles southwest of Mina. Property is inaccessible by automobile. Production has been 30 tons of scheelite ore.

Development consists of a tunnel 190 feet long, a shaft 48 feet deep, and some open-cuts totaling 500 feet.

These claims also are on the Silver Dyke vein system, and the values are in gold, silver, and tungsten, which are in a series of lenses.

Nevada Douglas Gold Mines, Inc.

The largest holdings in the Douglas area are owned by the Nevada Douglas Gold Mines, Inc., of which Harry E. Springer of Mina is the principal owner. Property comprises 16 patented and 4 unpatented claims.

Development includes a tunnel 500 feet long, a shaft at the tunnel portal, and other workings. In 1927 the company erected a 65-ton all-slime cyanide mill on the property which has operated for short periods. Mill equipment includes a Blake crusher, Hardinge ball mill, Dorr duplex classifier, Oliver filter, two Dorr agitation tanks, and three Dorr thickening tanks. The mill is operated by gasoline engine.

A tailings pile on the property estimated to contain 10,000 tons is reported to carry good values in gold.

The formation is mainly andesite, argillite, and dacite. A number of veins range from 1 to 5 feet in width. The principal gangue mineral is quartz, which is considerably shattered. Manganese and calcite also are present. The main vein which strikes nearly east and west, is at the contact of quartz porphyry and andesite. The width of this vein ranges from a few feet up to 30 feet. The smaller veins, from which most of the production has been derived, intersect the contact vein at acute angles. Values are chiefly in gold and silver.

High Ore Group

The High Ore group of three unpatented claims is owned by B. G. Strawser, 1733 Cardova

St., Los Angeles, Calif. In June 1936, the property was under lease to E. C. Van Allen, who worked with three men.

Development work consists of an incline shaft 130 feet deep and several hundred feet of lateral workings.

Equipment includes a Straub mill having a capacity of 2 1/2 tons in 9 hours and a gravity tram 600 feet in length from the mine to the mill. Ore is mined by hand methods.

In June 1936, a 7-day mill run produced 7 1/2 ounces of bullion worth $25 per ounce. The grade of the ore treated was $12.50 to $42 per ton.

Primary crushing is done with a 7- by 9-inch Blake-type crusher. After grinding in the Straub mill the values are recovered by amalgamation on a plate 7 1/2 feet long and 2 1/2 feet wide. By grinding to 30-mesh the recovery is about 75 percent of the values. The mill is operated by a 15- horsepower Stover gasoline engine.

Water for milling is hauled by truck from Pepper Springs 2 1/2 miles distant. Water consumption in milling is approximately 1,500 gallons per 9 hours. Part of the water is reclaimed.

The ore occurs in veins in rhyolite. Values are chiefly in gold.

Bentonite

A deposit of bentonite type of clay was discovered about 1928 on the east slope of Excelsior Mountains about 1 mile west of Sodaville.

Production of bentonite has been in the neighborhood of 15,000 tons from two claims owned by Cooper Shapley, formerly of Bishop, Calif. The bentonite was mined by power shovel and hauled by truck to Sodaville at a cost of 90 cents per ton. The bentonite was shipped to the Pacific coast markets for use as oil-well drilling mud. The deposit has been prospected by a number of trenches and shallow shafts.

Individuals living at Mina, Nev., who own bentonite claims are J. R. Towner, three; William Ray and William Gash, two, jointly; and George F. Thompson, several.

The bentonite occurs as a bedded deposit underlying surface detritus. It is said to be free from grit and in places is iron-stained and traversed by seams of gypsum. The overburden ranges in depth from 6 to 10 feet. Large reserves are indicated.

TEEL'S MARSH DISTRICT

Teel's Marsh is 2 miles south of the old mining camp of Marietta in southern Mineral County. It is reached by automobile road from Mina, a town on the Southern Pacific R. R. 26 miles to the northeast by way of Belleville.

This marsh, which in reality is a dry lake, was first worked for sodium chloride in the late sixties. The salt supplied the chlorination mills at Aurora, Comstock, and Candelaria. It is interesting to note that this marsh was the site of the first discovery of borax in Nevada by F. M. Smith, better known as "Borax" Smith, and J. P. Smith, his brother. According to S. T. Kelso of Hawthorne, Nev., who was at one time superintendent for the Smith Brothers, borax was found in Teel's Marsh about 1872. Shortly after, several plants for the extraction of borax were erected in the southeast portion of the marsh. These plants maintained a steady production up to 1892, when they were abandoned because of the discovery of richer deposits of the borax mineral, colemanite, in the vicinity of Death Valley, Calif. Although Teel's Marsh is not important economically at present, it produced a considerable quantity of borates and played an important part in the development of the borax industry in the United States.

Teel's Marsh is 5 miles long, 1 to 2 miles wide, and covers an area of about 8 square miles. The elevation of the surface of the marsh is 4,900 feet above sea level. The common salts in the playa deposits in the great basin region of which Teel's Marsh ia an example are the chlorides, sulphates, carbonates, bicarbonates, and borates of sodium and potassium. Magnesia and lime are present as minor constituents. In nearly all of the deposits the sodium salts predominate. Due to the fact that the borates are more soluble than the other salts, they remain in solution longer, and if the lake has alternate periods of desiccation and flooding the borates will crystallize out at or near the surface.

The deposition of borates in the playa type of deposits depends upon a combination of favorable conditions that is not widespread, and in consequence this type of deposit is restricted as to locality. The essential conditions required for the accumulation of borates are:

1. A source of boron, namely, solfataric springs in a region of former volcanic activity.
2. Suitable drainage basins, without any outlet, for the accumulation of the salts.
3. The climate must be sufficiently arid to concentrate the salts by evaporation and to prevent the removal of the borates, which are relatively soluble compounds.

At Teel's Marsh the boron-bearing mineral was principally borax, the natural sodium tetraborate intimately mixed with other salts forming a crust on the surface of the playa. The upper stratum of the deposits was the purest worked, but when this crust was removed other strata were found below at shallow depths associated with greater quantities of carbonate of soda or sodium chloride. The presence of sodium carbonate in the salts prevented the formation of the mineral ulexite nodules, the characteristic boron mineral in some dry-lake deposits.

The refining of the product was simple, as the natural borate of soda only required boiling to get the mineral in solution. On cooling, the borax was precipitated on wires or rods suspended in vats, leaving the other forms of soda, sand, and clayey matter at the bottom of the vat to be run off in the waste solution. The solution pans were semicircular, about 8 feet in diameter and 30 feet in length. These were fired from beneath with sage- brush, greasewood, or pinon pine from the foothills.

The crude borax obtained by crystallization was first.hauled with wagon teams to Wadsworth, Nev., 130 miles distant, for shipment by rail to the San Francisco Bay region, where refineries were located. In 1882 a narrow-gage railroad was completed to Mina, and long haul with a team was eliminated.

WHISKEY FLAT DISTRICT

The Whiskey Flat district is at the south end of Whiskey Flat, on the north slope of the Excelsior Mountains, about 20 miles southeast of Hawthorne. Mining was first done in this area in 1882, when copper ore carrying silver and a little gold was treated locally in a 400-pound-capacity furnace.

A company called the Excelsior Mountain Copper Co. operated for several years after 1907. Production has been small. There has been no activity in the district in recent years.

The mineralization is in limestone near a granite contact. In addition to silver and a little gold, the ore carries copper carbonates and sulphides. Garnet occurs as a gangue mineral.

FOOTNOTES

1. Statutes of the State of Nevada, 1911, pp. 10-15

2. Lotz, T. A., Report of the Surveyor General and State Land Register, State of Nevada, for the period Jan. 1, 1931, to June 30, 1932, incl.: State Printing Office, Carson City, Nev., p. 11.

3. Lincoln, Francis Church. Mining Districts and Mineral Resources of Nevada, pp. 137-157.

4. Lincoln., (work cited) p. 137. See also Statutes of the Territory of Nevada, 1861, pp. 50-53, 291.

5. Knopf, Adolph, Ore Deposits of Cedar Mountain, Mineral County, Nev.: U. S. Geol. Survey Bull. 725-H, 1933, 20 pp.

6. Director of the Mint, Report for 1883, p. 508.

7. Knopf, Adolph, The Candelaria Silver District, Nevada: U. S. Geol. Survey Bull. 735-A, 1922, 22 pp.

8. Knapp, S. A., Occurrence and Recovery of Sodium Carbonate in the Great Basin Region: Mineral Industry, vol. 7. 1898, pages 631-634.

9. Bentonite clay has the peculiar property of forming a colloid and absorbing several times its weight of water, swelling in the process to as much as 10 times its dry volume. Because of this property, it has been found suitable for sealing crevices in irrigating ditches, water reservoirs, and the like, where water is lost through seepage.

10. Hill, J. M., Some Mining Districts in Northeastern California and Northwestern Nevada: U. S. Geol. Survey Bull. 594, 1915, pp, 153-154.

11. Whitehill, H. R., Biennial Report of the State Mineralogist of the State of Nevada for years 1877-78-79, p. 25.

12. Whitehill, H. R., Biennial Report of the State Mineralogist: State of Nevada, 1875-76, p. 37.

13. Whitehill, H. R., Biennial Report of the State Mineralogist: State of Nevada, 1871-72, p. 38.

14. Whitehill, H. R., Biennial Report of the State Mineralogist: State of Nevada, 1873-74, p. 19.

15. Foshag, W. F., Quicksilver Deposits of the Pilot Mountains, Mineral County, Nev.: U. S. Geol. Survey Bull. 795. Oct. 27, 1929, pp. 113-123 (map).

16. Hess, Frank L.. and Larsen, Esper S., Contact-Metamorphic Tungsten Deposits of the United States: U. S. Geol. Survey Bull. 725-D, 1921, pp. 278-280.

17. Foshag, W. F., Work cited (footnote 15).

18. Foshag, W. F., work cited (footnote 15).

19. Amalgamation in pans heated by steam, using quicksilver, salt, and copper sulphate for reagents.

20. Burchard, H. C., Report of the Director of the Mint for 1882, p. 41.

21. Hill, J. M., Some Mining Districts in Northeastern California and North-western Nevada: U. S. Geol. Survey Bull. 594, 1915, 157-171.

22. Clark, C. W., Geology and Ore Deposits of the Santa Fe District, Mineral County, Nevada: Univ. of California Bull. Dept. Geol. Sci., vol. 14, no. 1, 1922, 74 pp.

23. A Sacramento mill is similar in design to an Ellis mill. Grinding is done with four iron balls, each weighing 145 pounds, rolled in a circular iron pan.

24. Work cited.

25. Kerr, Paul F., The Tungsten Mineralization at Silver Dyke, Nev.: Bull. University of Nevada, vol. 30, no. 5. 1936, 70 pp.

VIRGINIA CITY, SILVER REGION OF THE COMSTOCK LODE, by Douglas Mcdonald. Large 9x12 format, 128 pages, 75 illus., index. The discovery and development of the West's largest silver lode is recounted in an extensive text and both line drawing and photographs. Old and new maps help tell the story of this great mining town.

DEATH VALLEY GHOST TOWNS, VOL. II, by S.W. Paher. 32 pages, 9x12, illus. Mining camps of the Death Valley National Monument—Skidoo, Panamint City, and Old Stovepipe Wells—are joined by those immediately to the west, including Cerro Gordo, Darwin and Cartago.

GOLD IN THEM HILLS by C.B. Glasscock. Introduction by David F. Myrick. 330 pages, illustrated, map, index. Probably no newspaparman captured Nevada's wild 20th century mining boom better than Glasscock, and this is the finest of his six western books. This volume is dedicated to the burro; the prospector is the hero, but other mining camp followers round out the story. Chapters 2-4 tell the discovery and development of early Tonopah; chapters 5-17 discuss every important facet of Goldfield, including freighting, high-grading, the big mines, mining camp society and the fast-talking promoters. Death Valley's fabulous copper camp, Greenwater, occupies chapter 18, while the following chapter is the "birth and obituary" of Rhyolite. Gold in Them Hills will surely be enjoyed by all who love old Nevada.

LIFE IN THE GHOST CITY OF RHYOLITE, by Betsy Ritter. 64 pages. Here is a lively history of Rhyolite, the schools, the mines, the banks, the railroads, and founder Shorty Harris.

DESERT GEM TRAILS, by Mary Strong. 79 pages, maps. Here is a field guide to gems and minerals of southern California's Mojave and Colorado Deserts. Some sights in Goldfield area included.

Nevada's all-time best seller...

NEVADA GHOST TOWNS & MINING CAMPS, by S. W. Paher. Large 8-1/2x11 format, 492 pages, 700 illus., maps, index. About 668 ghost towns are described with directions on how to get to them. It contains more pictures and describes more localities than any other Nevada book. Nearly every page brings new information and unpublished photos of the towns, the mines, the people and early Nevada life. Now in 10th printing, it is the best seller of any book on Nevada ever published and it won the national "Award of Merit" for history. Cloth with color dust jacket.

GHOSTS OF THE GLORY TRAIL, by Nell Murbarger. 316 pages, illus., indexed. Subtitled "Intimate glimpses into the past and present of 275 western ghost towns," Ghosts of the Glory Trail is a fast-moving chronicle depicting the early-day mining stampedes. All Nevada counties are represented either in the 39 chapters on specific towns such as Aurora, Rhyolite, Candelaria, Hamilton, Unionville, Belmont, Tybo, El Dorado Canyon, Tuscarora, Delamar, etc.) or in the ghost town directory with 275 listings, some in California and Utah. Color cover.

RECREATIONAL GOLD PROSPECTING, by Jim Martin. 138 pages, illus. Basics of panning, use of rockers and cradles, dredges, claim staking, etc. Color cover.

COMSTOCK MINING & MINERS, by Eliot Lord. 451 pages, illus., maps. This comprehensive and well written narrative history of the Comstock Lode traces the birth of the silver mining industry in turbulent Virginia City right up to the original date of publication, 1883.

THE JOURNALS OF ALF DOTEN, by Alf Doten. 3 vols., 2381 pages, illus. 1870s journalist covers Comstock Lode during and after the boom years. Much on day to day living.

AN EDITOR ON THE COMSTOCK LODE, by Wells Drury. 343 pages, illus. these reminiscences of Comstock society comprise a vivid cross-section of life. Author portrays the bad men, the law and various personalities as he found them.

MINING CAMP DAYS by Emil W. Billeb. 229 pages, illus. The author provides insights into Nevada and eastern California mining camps after 1905. Dozens of unpublished photographs were taken by this observer-participant, augmenting a good text.

MARK TWAIN, YOUNG REPORTER IN VIRGINIA CITY, by Katharine Hillyer. 92 pages, illus. Informal collection of Twain episodes at the *Territorial Enterprise*.

SILVER KINGS: THE LIVES AND TIMES OF MACKAY, FAIR, FLOOD, AND O'BRIEN, by Oscar Lewis. 286 pages, illus. The story of the Comstock Lode is retold with short biographies of these four men who controlled the richest strike in North America.

HIGHGRADE, THE MINING STORY OF NATIONAL, NEVADA, by Nancy B. Shreier. 150 pages, illus. Detailed account of the rich National district north of Winnemucca, during 1907-13.

For information and prices write to:
Nevada Publications • Box 15444 • Las Vegas, Nevada

PLACER GOLD DEPOSITS IN NEVADA, by M.G. Johnson. (USGS Bull. 1356) 118 pages. County by county summary of 115 Nevada placer districts. Locations, extent of deposits and history.

THE MAKING OF A HARDROCK MINER, by Stephen Voynick. 224 pages, illus. Intimate description of life working underground—the heat, water, dark, constant danger. Much human interest.

VIRGINIA & TRUCKEE, A STORY OF VIRGINIA CITY AND THE COMSTOCK TIMES, by Lucius Beebe, 63 pages, illus., maps, biblio., tables. Few short lines were as familiar to powerful interests and celebrated to the world as the glorious V&T. This fast-moving account describes the mines that furnished ore for the silver mills along the Carson River. After 79 years the rails were ripped up in 1938.

NEVADA PLACE NAMES, by Helen Carlson. 280 pages. Treasure-trove of facts about the origins of names in Nevada's towns, natural features.

THE COMPLEAT NEVADA TRAVELER, by David W. toll. 192 pages, illus. The author divides Nevada into five regions in order to describe in each of them its history, services for travelers, annual events and assorted trivia. Interesting photographs (some in color) are interspersed amid a lively text.

TOURING NEVADA: A HISTORIC AND SCENIC GUIDE, by Mary Ellen and Al Glass. 253 pages, illus. Nevada travel descriptions for visitor and resident alike. Maps and color index are definite aids. Color cover.

GEOLOGY OF THE GREAT BASIN, by Bill Fiero. 250 pages, illus. A fine treatment complete with maps, charts, and analysis of Nevada geology.

THE STORY OF CANDELARIA AND ITS NEIGHBORS: COLUMBUS, METALLIC CITY, BELLEVILLE, MARIETTA, SODAVILLE AND COALDALE, by Hugh Shamberger. 200 pages, illus. History of mining in southwestern Nevada; an important study.

COPPER TIMES, by Jack Fleming. 255 pages, Collection of stories on Whte Pine County—its people, towns, copper industry, recreation, education, history.

EUREKA AND ITS RESOURCES, by Lambert Molinelli. 136 pages, illus. Good, short history of the mines, issued in 1879.

RAWHIDE, by Hugh A. Shamberger. 50 pages, maps. Early history, development, and water supply of this mining camp in Mineral County.

THE TOWN THAT DIED LAUGHING, by Oscar Lewis. 235 pages. The story of Austin, a rambunctious early day Nevada mining camp, and of its newspaper, The Reese River Reveille.

A MINER'S CHRISTMAS CAROL, AND OTHER FROMTIER TALES, by Sam Davis. 86 pages. A turn-of-the-century Carson City editor writes of early Pioche, Carson City, Virginia City, and newspapering.

THE NEWSPAPERS OF NEVADA, by Richard Lingenfelter. 336 pages, illus., index. History of 800 publications issued from various Nevada localities. A thorough study.

RAILROADS OF NEVADA AND EASTERN CALIFORNIA, VOL. 1, by David F. Myrick. 451 pages, illus, maps. Author details 43 Nevada lines, from the Central Pacific to those serving Tonopah-Goldfield-Ely booms. Numerous illustrations augment the definitive text which also includes politics, social life, and Nevada personalities.

DEEP ENOUGH, by Frank Crampton. 281 pages, True-to-life autobiography of a working stiff in the Nevada mining camps.

INDIANS OF COO-YE-EE PAH, by Nellie Harner. 145 pages. History of Northern Paiutes from a native's point of view.

REPORT OF THE EXPLORATIONS ACROSS THE GREAT BASIN IN 1859 FOR A DIRECT WAGON-ROUTE TO GENOA, by James Simpson. 518 pages, illus. Author moved westward across central Nevada in 1858, noting the plants, animals, birds, Indians, etc.

GOLDFIELD, by Hugh A. Shamberger. 240 pages, illus. The early history, the mines, the struggle for water, the building of railroads and the Gans-Nelson fight of 1906—are described. Author details rise of corporate mining and highgrading.

Write for catalog
of more than 100 ghost town
and treasure books
on the
Southwestern states.

Out-of-print Books

We search for out-of-print books—send us your want lists.

Nevada Publications • Box 15444 • Las Vegas, Nevada

Nevada Publications • Box 15444 • Las Vegas, Nevada

U.S. Geological Survey Mining Bulletins Reprint Series

These bulletins, issued early in this century, are much in demand as primary reference material for Nevada mining and water sources. All cover extensively mineralogy, topography, history and the various mines. All are 6 x 9 quality paperbacks with pages sewn together.

MINES OF THE GOLDFIELD, BULLFROG AND OTHER SOUTHERN NEVADA DISTRICTS, by F. L. Ransome, 144 pp, maps, index. Color map 22 x 28 laid in. This 1907 Bulletin also covers Searchlight, Eldorado, Crescent and Gold Mountain, The last 42 pages reproduce from Out West magazine two contemporary illustrated articles. "The Nevada Bonanzas of Today" and "Goldfield and the Goldfielders." These capture the flavor and excitement of Nevada's mining era.

MINES OF THE SILVER PEAK RANGE, KAWICH RANGE AND OTHER SOUTHWESTERN NEVADA DISTRICTS, by S.H. Ball. 218 pp, maps, index. Color map 22 x 28 laid in. This 1907 Bulletin also covers Lida-Goldfield Valley, Pahute Mesa, Cactus Range, Reveille Range, Belted Range, Tolicha Peak, the Amargosa Mountain system, Gold Mountain, Death Valley and the Panamint Range.

320 DESERT WATERING PLACES IN SOUTHEASTERN CALIFORNIA AND SOUTHWESTERN NEVADA, by W. C. Mendenhall. 104 pp, illus, index. Color map 14 x 20 laid in. This 1909 Water-Supply Paper covers the Death Valley Basin in Inyo County and the Salton Sink of Imperial County, and numerous places in between. There are descriptions of 323 springs, all numbered and keyed to the colored map. The bulk of the book concerns San Bernardino County; in Nevada the following counties: Esmeralda, Nye, and Clark.

MINES OF BATTLE MOUNTAIN, REESE RIVER, AURORA AND OTHER WESTERN NEVADA DISTRICTS, by J. M. Hill. 200 pp, maps, index. This 1915 Bulletin also covers California's Lassen and Modoc counties, Douglas County, the Peavine district, and in Mineral County the Hawthorne, Granite, Santa Fe, Silver Star, and Pine Grove in Lyon County.

MINES OF TUSCARORA, CORTEZ AND OTHER NORTHERN NEVADA DISTRICTS, by W. H. Emmons. 220 pp, maps, index. This 1910 Bulletin also covers Midas, Good Hope, Columbia, Edgemont, Mountain City, Bullion, Safford, Tenabo, Lander, and the Lewis districts. Included is F. L. Ransome's "Notes on Some Mining Districts in Humboldt County" [Pershing County] 1909. Districts included are: Seven Troughs, Rosebud, Red Butte, Humboldt Range, Pahute Range, Sonoma Range.

MINES OF CHERRY CREEK, BRISTOL AND OTHER EASTERN NEVADA DISTRICTS, by J. M. Hill. 214 pp, maps, index. This 1916 Bulletin also includes Gold Butte, the Ruby Range, Mud Springs, Spruce Mountain, Mizpah, Kinsley, Toano Range, Tecoma, Ravenswood, Atlanta, Patterson, Troy, Bald Mountain, Granite, Hunter, Ward, Aurum, Duck Creek, Taylor and Kern districts.

An Essential Mining Reference...

MINING DISTRICTS AND MINERAL RESOURCES OF NEVADA, by Francis C. Lincoln. 295 pp, maps, index. This compilation gives a summary of each mining district in Nevada through 1923. There are historical summaries and bibliographies of each mining area, all arranged by counties. Section Two describes Nevada's mineral resources by types. An essential tool for the mining researcher.

NEVADA MINING REPRINTS
by William O. Vanderburg

All feature information on past operations, nature of mineral deposits and geology.

MINES OF CLARK COUNTY, 96 pp, illus, map. This report originally issued in 1937, covers many southern Nevada districts: Alunite, Arden, Crescent, Eldorado Canyon, Gold Butte, Goodsprings, St. Thomas, Searchlight.

MINES OF HUMBOLDT AND PERSHING COUNTIES, 128 pp, illus, map. This report, originally published in 1936 and 1938, covers many north-central Nevada districts: Awakening, Dutch Flat, Golconda, Gold Run, McDermitt, National, Sulphur, Warm Springs, Arabia, Farrell, Gold Banks, Kennedy, Mill City, Placerites, Rabbit Hole, Rochester, Rosebud, Scossa, Seven Troughs, Star, Trinity, Unionville, and others.

MINES OF LANDER AND EUREKA COUNTIES, 160 pp, illus, map. This report originally issued in 1938 and 1939, covers many eastern Nevada districts: Battle Mountain, Big Creek, Bullion, Hilltop, Kingston, Lewis, McCoy, Reese River (Austin), Buckhorn, Cortez, Diamond, Eureka, Mineral Hill, Palisade, Union, and others.

MINES OF CHURCHILL AND MINERAL COUNTIES, 144 pp, illus, map. This report, originally issued in 1937 and 1940, covers many west-central Nevada districts: Bernice, Fairview, Jessup, La Plata, Wonder, Aurora, Broken Hills, Candelaria, Garfield, Marietta, Rawhide, Silver Star and others.

CAMELS IN NEVADA, by Douglas McDonald. 32 pages, illus. In Nevada camel pack trains hauled salt, wood and even freight, also aiding early surveyors. But the beasts also brought problems. Modern camel races in Virginia City are recounted. Color cover.

THE BIG BONANZA, by Dan DeQuille (William Wright). 488 pages, illus., with intro. by Oscar Lewis. Indexed. Subtitled "An authentic account of the discovery, history, and working of the Comstock Lode," the Big Bonanza by DeQuille covers every phase of the epic rise of Virginia City, especially the special technology required to work the deep silver mines. Color cover.

LOST LEGENDS OF THE SILVER STATE, by Gerald Higgs. 142 pages, illus., Obvious, better known Nevada stories are dutifully ignored. Included are stories of Mark Twain being thrown out of Nevada, Civil War days, the $100.00 boulder, etc.

HISTORY OF NEVADA, 1913, by Sam P. Davis. 2 vols. 1344 pages, 60 illus. Originally issued in 1913, this landmark history is a compilation of special treatises on Nevada geography, Indians, territorial life, law and crime, mining history, politics, and biographies of prominent early 20th century Nevadans. The new index has 6,000 subject entries.

NEVADA TOWNS & TALES, by S. W. Paher, ed. 2 vols. 224 pages ea. 8-1/2x11. Chapters focus on Nevada's economic, social and geographical factors. Other major sections discuss state emblems, gambling, politics, mining, business and casino entertainment. There is much material on ghost towns, prospecting, legends, early day women, ranching, native animals, industries, banking and commerce, railroads, atomic testing, transportation, etc. Indexed. Color cover. Vol. 1, North, Vol. 2, South.

MY ADVENTURES WITH YOUR MONEY, by George Graham Rice. 334 pages, illus. Here are the memoirs of get-rich-quick financing of central Nevada and Death Valley mines, with interesting anecdotal material. Author capitalized the stocks of Goldfield, Greenwater and Rawhide mines, listed them on the national exchanges, and reaped the profits until convicted of mail fraud in 1911. More than 110 illustrations complement the text.

VIRGINIA CITY'S INTERNATIONAL HOTEL, by Richard C. Datin. 49 pages, illus. Subtitled Elegance on C Street. This book recreates the history, the people and the splendor of Virginia City. The International Hotel typified it all, kings, financiers, President U.S. Grant, and queens of the footlights.

SKETCHES OF VIRGINIA CITY, N.T., by J. Ross Browne. 48 pages, illus. In 1860 the author commented extensively on the miners and their madness over minerals, the Chinese, the Indians, the stagecoach drivers, proprietors, barroom brawlers, etc. Charming, humorous cartoons of these appear in the book.

NEVADA, AN ANNOTATED BIBLIOGRAPHY, by S.W. Paher. 585 pages, 7x10 illus. Here is a researcher's guide and description of more than 2,544 books relating to the history and development of Nevada. The useful 74-page index has 3,550 subjects; listed are 33 other bibliographies referring the researcher to countless Nevada materials.

JULIA BULETTE AND THE RED LIGHT LADIES OF NEVADA, by Douglas McDonald. 32 pages, illus., map. Here is the best written historical sketch to date of Virginia City's famed prostitute who was murdered in 1867. An overview of Nevada prostitution occupies the last part of the book, augmented by interesting photographs.

BODIE...BOOM TOWN. GOLD TOWN! The last of California's Old-time Mining Camps by Douglas McDonald. 48 pages, illustrated, color cover, 7x10 format. Though Bodie was discovered in 1859, no significant mining was begun until phenomenally rich strikes were made in 1877. The height of the boom occurred during 1879-80, though the mines were still active until about 1920. This heavily illustrated book shows the mines and miners, street scenes, important buildings, the mill, bullion, and the crowds which made up this great old mining camp.

MARK TWAIN IN VIRGINIA CITY, NEVADA, by Mark Twain. 192 pages. Twain portrays the life and the people of Virginia City, including the mining litigation, breaking in of a horse, etc. He mined for silver, labored in a silver mill and worked as a reporter for Virginia City's Territorial Enterprise. Drawings and cartoons. Paper.

For information and prices write to:

Nevada Publications • Box 15444 • Las Vegas, Nevada

Remember these books for birthday gifts, at Christmas, Mother's Day and Father's Day